Zero and Net Zero Energy

Edited by Getu Hailu

Published in London, United Kingdom

IntechOpen

Supporting open minds since 2005

Zero and Net Zero Energy
http://dx.doi.org/10.5772/intechopen.78872
Edited by Getu Hailu

Contributors
Samuel I. Egwunatum, Ovie I. Akpokodje, Yan Liu, Liu Yang, Yuhao Qiao, Jiang Liu, Mengyuan Wang, Jesús Cerezo Román, Antonio Rodríguez Martínez, Rosenberg Javier Romero Domínguez, Pedro Soto Parra, Ahmed A.Y. Freewan, Bohumír Garlík, Ruta Miniotaite, Getu Hailu

First published in London, United Kingdom, 2019 by IntechOpen
IntechOpen is the global imprint of INTECHOPEN LIMITED, registered in England and Wales, registration number: 11086078, 7th floor, 10 Lower Thames Street, London, EC3R 6AF, United Kingdom
Printed in Croatia

British Library Cataloguing-in-Publication Data
A catalogue record for this book is available from the British Library

Additional hard and PDF copies can be obtained from orders@intechopen.com

Zero and Net Zero Energy
Edited by Getu Hailu
p. cm.
Print ISBN 978-1-78984-498-6
Online ISBN 978-1-78984-499-3
eBook (PDF) ISBN 978-1-83881-881-4

For EU product safety concerns:
IN TECH d.o.o., Prolaz Marije Krucifikse Kozulić 3, 51000 Rijeka, Croatia, info@intechopen.com or visit our website at intechopen.com.

Meet the editor

Dr. Getu Hailu is currently an Assistant Professor of mechanical engineering at the University of Alaska Anchorage. He has more than 20 years of experience in research and teaching. His current research focus is on technologies for net zero energy buildings (including solar thermal storages, building integrated photovoltaic thermal systems, solar energy and air source heat pump application in cold regions). His teaching interests include courses in the thermofluids and renewable energy areas both at the graduate and undergraduate level. He is the author or coauthor of more than 40 refereed publications. He is a member of ASME, ASHRAE and IBPSA.

Contents

Preface

The definition of zero-energy buildings (ZEB) and net-zero energy buildings is somewhat unclear. According to the National Renewable Energy Laboratory (NERL), at the heart of the ZEB concept is the idea that buildings can meet all their energy requirements from low-cost, locally available, nonpolluting, and renewable sources. At the strictest level, a ZEB generates enough renewable energy on-site to equal or exceed its annual energy use. According to the US Department of Energy (DOE), a ZEB is a building that produces enough renewable energy to meet its own annual energy consumption requirements, thereby reducing the use of non-renewable energy in the building sector. DOE further states that ZEBs use all cost-effective measures to reduce energy usage through energy efficiency and include renewable energy systems that produce enough energy to meet remaining energy needs. According to DOE, advantages of ZEBs include lower environmental impacts, lower operating and maintenance costs, better resiliency to power outages and natural disasters, and improved energy security.

Setting aside the ambiguity in the definition, there is a growing concern about fluctuating energy prices, energy security, and the impact of climate change. Buildings are amongst the primary energy consumers worldwide. This fact underlines the importance of targeting building energy use as key to decreasing any nation's energy consumption. According to the American Society of Heating, Refrigeration and Air Conditioning Engineers (ASHRAE) Research Strategic Plan 2010-2015, even limited deployment of Net-Zero-Energy buildings within this timeframe will have a beneficial effect by reducing the pressure for additional energy and power supply and the reduction of GHG emissions. The building sector is poised to significantly reduce energy use by incorporating energy-efficient strategies into the design, construction, and operation of new buildings and retrofits to improve the efficiency of existing buildings. The building sector can substantially reduce dependence on energy derived from fossil fuels by increasing use of on-site and off-site renewable energy sources.

The book has seven chapters, which are divided into two sections: Zero Energy Buildings, and Economic Prospects of Zero Energy Buildings. As the section names indicate, the book provides some technical and economic aspects of ZEBs. The book is useful as a reference for students, practicing engineers, and general public.

Dr. Getu Hailu
University of Alaska Anchorage,
Anchorage, Alaska, USA

Section 1

Introduction

Chapter 1

Introductory Chapter: Path to Net Zero Energy Buildings

Getu Hailu

1. Overview

Energy demand and usage is expected to change significantly with changing weather patterns, affecting heating/cooling demands and electricity demands. Energy supplies will face changing conditions, such as reduced efficiency of thermal plants, cooling constraints on thermal plants, and increased pressure on transmission and distribution systems. International Energy Agency (IEA) estimates 1°C of temperature increase can reduce the available summer electricity generation capacity up to 16% by 2040 in the United States alone [1]. Sea level rise, permafrost melting, intense and more frequent extreme weather events, increased wind speeds, and ocean storms will all negatively impact energy infrastructure. For example, large numbers of overhead power lines over extended distances could easily be brought down. Consequently, it is likely that the building sector will be highly impacted by climate change and associated weather patterns. It is also true that the building sector is well positioned and has the potential to mitigate such effects to a great extent.

According to the American Society of Heating, Refrigerating and Air-Conditioning Engineers (ASHRAE) Research Strategic Plan 2010–2015, even limited deployment of net zero energy (NZE) buildings within this timeframe will have a beneficial effect by reducing the pressure for additional energy and power supply and the reduction of greenhouse gas (GHG) emissions [2]. The implementation of NZE buildings requires use of multiple innovative technologies and control strategies for space heating and cooling and water heating. Hybrid photovoltaic-thermal (PV/T) systems, building-integrated photovoltaics (BIPV), and thermal energy storages have been identified by the US Department of Energy (DOE) as technologies that could make substantial contributions toward that goal [3].

2. Current state of the art

Attempts have been made in using distributed energy systems (DRE) to meet electricity and thermal energy demand of a building. For example, photovoltaic/thermal (PV/T) systems, which produce electricity and heat, have been applied effectively to building roofs and facades to offset or eliminate fossil fuel demand in buildings. But they are treated as separate and distinct systems from each other and from the building envelope. This lack of system integration represents a lost opportunity to simplify and derive additional gains in efficiency. To address this PV/T system integration into the building structure has been the next step in research and development. Arrays of photovoltaic modules with heat recovery capability, which are integrated into the building envelope so that the assembly replaces elements of the facade and/or the roof, thereby reducing overall cost, are known as building-integrated photovoltaic/thermal (BIPV/T) systems. It has been reported that BIPV/T systems have the advantages of (1) reducing the temperature of the PV panels and increasing electrical efficiency, (2) extending the life of the system by reducing the tendency of the modules to delaminate, and (3) recovering thermal energy for spacing heating and domestic hot water heating purposes. BIPV/T systems have the potential to meet all building envelope requirements, such as mechanical resistance and thermal insulation [4–7]. Heat recovery is accomplished by fluid circulation behind the array of photovoltaic modules, which heat up when exposed to sunlight. This thermal energy either can be directly used for space and/or domestic water heating or can be delivered to an air source heat pump (ASHP) to enhance its performance [6–10]. This is particularly important where the coefficient of performance (COP) of the ASHP decreases because of colder outdoor temperatures. It has been reported that ASHPs offer low initial cost compared to ground source heat pumps (GSHP), with 40% reduction in installation cost [4], but their COP decreases in colder outdoor temperatures [11]. Coupling ASHPs with BIPV/T systems has been reported to have the potential to further reduce building heating and cooling costs and dependence on nonrenewable heating fuels. For example, it has been reported that the energy consumption of the ÉcoTerra (Montreal) house was only 26.8% of a typical Canadian home when a BIPV/T system was coupled with a GSHP [12]. As mentioned earlier, with a 40% reduction in installation cost [4], and with a "pre-treatment" of the outdoor air by a BIPV/T system, ASHPs coupled to a BIPV/T system are an attractive alternative to GHSPs coupled to BIPV/T systems and have great potential to increase building energy efficiency.

Another technology that has been shown to improve the thermal performance of buildings significantly is thermal energy storage (TES) system [13–17]. This is especially attractive in regions with extended period of freezing, such as Alaska, where the performance of ASHPs is poor during cold weather. Concrete is the most common and effective building material used as thermal mass [18]. Concrete slabs can be utilized as effective TES systems [19–21]. It has been suggested that since the BIPV/T system is an air-based system, the collected thermal energy can be released to concrete slab TES during daytime. During the night, when the outdoor temperature is cold, the stored thermal energy can be released to the air source heat pump, leading to the enhancement of the overall coefficient of performance of the ASHP.

Renewable energy sources often are not available when needed or do not meet the fluctuating demand for heating. For example, in summer, a thermal solar heater's output is large compared to winter time. In such cases, seasonal thermal storage systems are employed. In fact, energy storage is a key to facilitating the widespread use of many renewable energy resources. Large heating and cooling loads can be addressed through seasonal solar thermal energy storage—SSTES [22]. Thermal energy storage is also a key element in building mechanical systems. It allows covering heating and cooling needs in an efficient and economical way, particularly domestic hot water—DHW [23]. Sensible heat storage, using readily available simple materials such as rocks, is one of the most widely used techniques for thermal energy storage. It is simple and least expensive [24, 25]. In addition, sensible heat storage systems have the advantage of reversible charging and discharging capabilities for unlimited number of cycles, i.e., over the life span of the storage, unlike, for example, phase change materials [26]. The use of simple materials, such as rocks and sand which are readily available in many areas, makes sensible heat storage long lasting, safe and relatively easy to install, and applicable for remote areas [25]. Sensible heat storage systems have applications in residential, industrial, and commercial settings. For example, the Drake Landing Solar Community uses a combination of seasonal ground-based thermal storage with short-term liquid storage tanks [27]. Large-scale seasonal storage systems have been constructed in Switzerland, Denmark, Finland, France, the Netherlands, the United States, Turkey, Korea, Germany, and Canada.

Another thermal storage technology, latent heat storage (LHS) systems, involves the storage of energy in phase change materials (PCMs). Thermal energy is stored and released with changes in the material's phase. LHS has the advantage of being compact, i.e., for a given amount of heat storage, the volume of PCMs is significantly less than the volume of sensible heat storage. This allows for less insulation material and applicability in places where space is limited. Another advantage of PCM is that they can be applied where there is a strict working temperature, as the storage can work under isothermal conditions. Phase-transition enthalpy of PCMs is usually much higher (100–200 times) than sensible heat. Consequently, latent heat storages have much higher storage density than sensible heat storages [28]. Current PCM research is mainly focused on technologies that deal with materials (i.e., storage media for different temperature ranges), containers, and thermal insulation development. There is still a need for basic and applied research and development of design methods, research and development to improve performance analysis; reduce operation cost of installed systems; assure their long-term, smooth operation; and improve their efficiency. More research is also required in understanding system integration and process parameters as well as improving reacting materials [29].

Another technology that stores thermal energy from the sun is in the form of chemical energy. The process is known as solar thermochemical energy storage (TCES). Through a reversible endothermic chemical reaction, energy is stored as a chemical potential using the solar thermal energy. TCES does not yet show clear advantages for building applications, despite the potentially high energy density. Currently, there is no available material for thermochemical energy storage that satisfies all the requirements for building operations. Besides, thermochemical solutions require different tanks and heat exchangers that should be carefully addressed for small-scale applications. Also, additional research efforts are needed to optimize operation conditions, efficiency, costs, and system designs.

In conclusion, the path to zero and net zero energy buildings requires multiple technologies working together. Currently, these systems are treated as separate and distinct systems from each other and from the building envelope. There is a need for system integration and development of effective control strategies to simplify and derive additional gains in building energy efficiency. Buildings with integrated distributed energy systems are resilient to catastrophic weather events.

Author details

Getu Hailu
Department of Mechanical Engineering, University of Alaska Anchorage, USA

*Address all correspondence to: ghailu@alaska.edu

References

[1] Annual Energy Outlook 2016 Early Release: U.S. Energy Information Administration. Available from: https://www.eia.gov/outlooks/aeo/er/ [Accessed November 2017]

[2] ASHRAE research strategic plan 2010-2015

[3] Goetzler W, Guernsey M, Droesch M. Research and Development Needs for Building Integrated Solar Technologies. U.S. Department of Energy Office of Energy Efficiency and Renewable Energy Building Technologies Office Report; 2014

[4] Safa A, Fung AS, Kumar R. Performance of a two-stage variable capacity air source heat pump: Field performance results and TRNSYS simulation. Energy and Buildings. 2015;**94**:80-90

[5] Chen Y, Athienitis AK, Galal KE. Modeling, design and thermal performance of a BIPV/T system thermally coupled with a ventilated concrete slab in a low energy solar house: Part 2, ventilated concrete slab. Solar Energy. 2010;**84**(11):1908-1919

[6] Getu H, Dash P, Fung AS. Performance evaluation of an air source heat pump coupled with a building integrated photovoltaic/thermal (BIPV/T system). In: ASME 2014 8th International Conference on Energy Sustainability; Boston, Massachusetts, USA. p. 6

[7] Getu H, Yang T, Athienitis AK, Fung A. Computational fluid dynamics (CFD) analysis of air based building integrated photovoltaic/thermal (BIPV/T) systems for efficient performance. eSIM; 7-10 May 2014; Ottawa, Canada; 2014

[8] Getu H, Dash P, Fung A. Performance evaluation of an air source heat pump coupled with a building-integrated photovoltaic/thermal (BIPV/T) system under cold climatic conditions. Energy Procedia. 2015;**78**:1913-1918

[9] Roeleveld D, Getu H, Fung AS, Naylor D, Yang T, Athienitis AK. Validation of computational fluid dynamics (CFD) model of a building integrated photovoltaic/thermal (BIPV/T) system. Energy Procedia. 2015;**78**:1901-1906

[10] Kamel R, Ekrami N, Dash P, Fung AS, Getu H. BIPV/T + ASHP: Technologies for NZEBs. Energy Procedia. 2015;**78**:424-429

[11] Bertsch S, Groll E. Two-stage air-source heat pump for residential heating and cooling applications in northern U.S climates. International Journal of Refrigeration. 2008;**31**(7):1282-1292

[12] Doiron M, O'Brien W, Athienitis AA. Energy performance, comfort and lessons learned from a near net-zero energy solar house. In: ASHRAE Transactions 117; 2011. pp. 585-96

[13] Chen Y, Athienitis AK, Galal K. Thermal performance and charge control strategy of a ventilated concrete slab (VCS) with active cooling and using outdoor air. ASHRAE Transactions 118; 2012. pp. 556-568

[14] Dincer I. On thermal energy storage systems and applications in buildings. Energy and Buildings. 2013;**34**(4):377-388

[15] Hadorn JC, editor. Thermal Energy Storage for Solar and Low Energy Buildings: International Energy Agency (IEA) Solar Heating and Cooling Task 32—Advanced Storage Concepts for Solar and Low Energy Buildings, IEA; 2005

[16] Howard B. The CMU air-core passive hybrid heat storage system. In: Proceedings of the Renewable and Advanced Energy Systems for the 21st Century; 11-15 April 1999

[17] Morgan S, Krarti M. Impact of electricity rate structures on energy cost savings of pre-cooling controls for office buildings. Building and Environment. 2007;**42**(8):2810-2818

[18] ACI Committee 122, 122R-02. Guide to thermal properties of concrete and masonry systems. In: American Concrete Institute (ACI), ed. Manual of Concrete Practice; Detroit, Mich, US: American Concrete Institute; 2002

[19] Shaw MR, Treadaway KW, Willis STP. Effective use of building mass. Renewable Energy. 1994;**5**(2):1028-1038

[20] Feustel HE, Stetiu C. Hydronic radiant cooling—Preliminary assessment. Energy and Buildings. 1995;**22**(3):193-205

[21] Inard C, Meslem A, Depecker P. Energy consumption and thermal comfort in dwelling-cells: A zonal-model approach. Building and Environment. 1998;**33**(5):279-291

[22] Vadiee A, Martin V. Thermal energy storage strategies for effective closed greenhouse design. Applied Energy. 2013;**109**:337-343. DOI: 10.1016/j.apenergy.2012.12.065

[23] Athienitis A, O'Brien W. Modeling, Design, and Optimization of Net-Zero Energy Buildings. Berlin, DE: Wilhelm Ernst & Sohn Verlag fur Architektur und Technische; 2015

[24] Tatsidjodoung P et al. A review of potential materials for thermal energy storage in building applications. Renewable and Sustainable Energy Reviews. 2013;**18**:327-349. DOI: 10.1016/j.rser.2012.10.025

[25] Pinel P, Cynthia AC, Ian B-M, Adam W. A review of available methods for seasonal storage of solar thermal energy in residential applications. Renewable and Sustainable Energy Reviews. Elsevier; 2011;**15**(7):3341-3359. DOI: 10.1016/j.rser.2011.04.013

[26] Rosen MA. Net-zero energy buildings and communities: Potential and the role of energy storage. Journal of Power and Energy Engineering. 2015;**3**:470-474. DOI: 10.4236/jpee.2015.34065

[27] Garg HP, Mullick SC, Bhargava AK. Solar Thermal Energy Storage. Boston, US: Springer; 1985. pp. 181-212

[28] Tian Y, Zhao CY. A review of solar collectors and thermal energy storage in solar thermal applications. Applied Energy. 2013;**104**:538-553

[29] Shah YT. Thermal Energy: Sources, Recovery, and Applications. Boca Raton, FL, USA: CRC Press; 2018

Section 2

Towards Zero Energy Buildings

Chapter 2

Building Envelope with Phase Change Materials

Liu Yang, Yan Liu, Yuhao Qiao, Jiang Liu and Mengyuan Wang

Abstract

Based on recent investigations on building envelope with phase change materials from all over the world, we select the key scientific and technical issues including the thermal design methods, climatic and seasonal suitability and application, etc. The chapter mainly contains four parts: how to design building envelope with phase change materials, how to deal with issues on climatic and seasonal suitability of the technology, how to improve thermal performance of phase change materials applied in building envelope, and what is the application mode. The thermal design principle and a simple calculation method of building envelope with phase change materials are proposed by experiments. Thermal comfort pertaining to ASHRAE Standard 55 under different conditions is investigated, and an approach to estimate favorable climatic characteristics for building envelope with phase change materials is established. To exert the phase change materials applied in building envelope effectively, thermal transfer enhancement methods and application are also provided in the chapter. The chapter can be helpful for the development of building energy efficiency and the goal of zero and net zero energy.

Keywords: building envelope, phase change materials, thermal design, climatic and seasonal suitability, application

1. Introduction

Building energy saving is essential to overall energy conservation from different sectors [1]. To build a comfortable indoor thermal environment, the energy consumption of air conditioning is increasing rapidly, which has negative impacts on sustainable development. Passive low-energy buildings are developed to solve the problem [2]. Improving thermal performance of building envelope is an effective approach to achieve a stable indoor thermal environment and reduce building energy consumption [3]. There are two main ways for the improvement of building envelope thermal performance. One is to reduce heat transfer coefficient (U-value) then decrease the heat flux of building envelope. Another is to increase thermal inertia of buildings to enhance the resistance to the changing of outdoor thermal environment [4, 5], especially in climate conditions with large daily temperature range, where remarkable energy efficiency performance could be achieved by improving building thermal stability [6, 7]. Adopting heavy structure (such as earth brick) is a traditional method for the improvement of building thermal stability [8]. However, the traditional approach is not suitable for achievement of satisfying

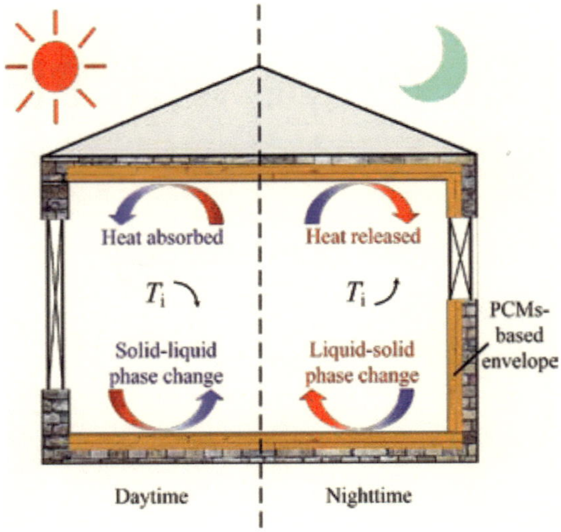

Figure 1.
Principle of PCM-based envelope.

thermal storage performance in lightweight structures widely developing in recent years. Due to large thermal capacity, phase change materials (PCMs) are ideal thermal storage materials to be integrated with building envelope [9]. As shown in **Figure 1**, building envelope integrated with PCMs can improve thermal inertia of lightweight structures and further stabilize indoor thermal environment [10]. Therefore, the thermal performance of PCM-based envelope has attracted more attention in recent years.

2. How to design building envelope with phase change materials

In this part, a kind of PCM-based lightweight wallboards which integrates PCMs with insulation materials is put forward. In application, the different PCM layer arrangements and the different PCM layout areas are significant to thermal performance of the wallboards. Current thermal design calculation of PCM-based envelope is mainly based on numerical methods. Although the numerical methods could obtain the wall temperature accurately, it is difficult to provide amounts of parameters and build models for architects. Therefore, in order to provide reference for the estimation of PCM-based envelope application effects, a thermal design calculation method is proposed, which is based on harmonic response method and equivalent specific heat capacity principle and is verified with experimental results. The present part mainly contains two points as follows: (1) a comparative experimental investigation with reduce-scale in the controlled condition, and (2) a simple thermal design method is developed based on harmonic response method and equivalent specific heat capacity principle.

2.1 Reduce-scale experiments

The experiments are conducted in the Thermal Storage and Ventilation Laboratory (TSVL) in Xi'an University of Architecture and Technology, Shaanxi, China. The lab consists of artificial climate chamber and control system, which can simulate the outdoor thermal environment including temperature and humidity. The temperature can be controlled by setting curves. Two test cells are adopted in the comparative experiments. The length, width, and height in each test cell are 1200mm, 660mm, and 800 mm, respectively. To exclude the uncontrollable factor influence, the wallboards of test cell are made by 100 mm expandable polystyrene (EPS) panels which can be considered as adiabatic. In the experiments, PCM-based lightweight wallboard is adopted to replace one of EPS panel (see **Figure 2**). The macro-encapsulated PCM panels are employed in the experiments. The encapsulated PCMs are $CaCl_2{\cdot}6H_2O$, whose phase change temperature is 25–27°C. In melting process, the latent heat is 122.3 $kJ{\cdot}kg^{-1}$, and in solidification process the latent heat is 116.9 $kJ{\cdot}kg^{-1}$.

The experimental wallboard named PCM-based lightweight wallboard is put forward. The wallboard is made by 20 mm PCM layer and 30 mm insulation material layer. The wallboard improves its thermal storage capacity obviously by utilizing PCMs while sacrificing a fraction of insulation performance. The PCM-based lightweight wallboard is divided into PCM part and insulation part to distinguish whether there are PCMs in the horizontal direction.

r_A is defined as the ratio of PCM part area to wallboard area:

$$r_A = \frac{A_{PCM}}{A_{ALL}} \tag{1}$$

where A_{PCM} is the area of PCM and A_{ALL} is the area of the wallboard. The wallboards with four r_A are compared in experiments: 0, 21.6, 43.3, and 65%.

The reference group (r_A = 0%) is performed firstly to benchmark the PCM-based lightweight wallboard experiments. **Figure 3** shows the temperature curves of reference group during harmonic temperature changing process. Comparing the

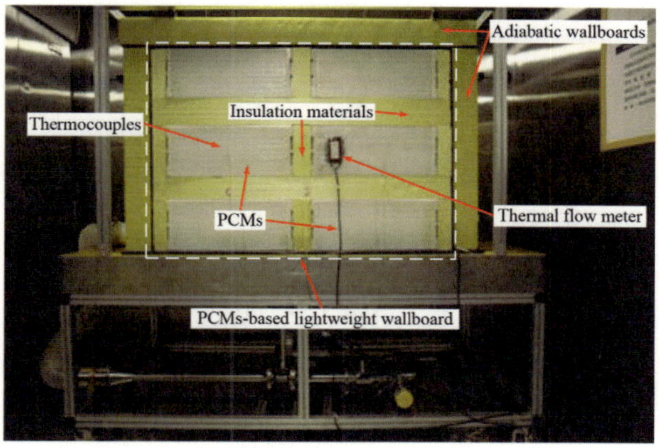

Figure 2.
PCM-based lightweight wallboard.

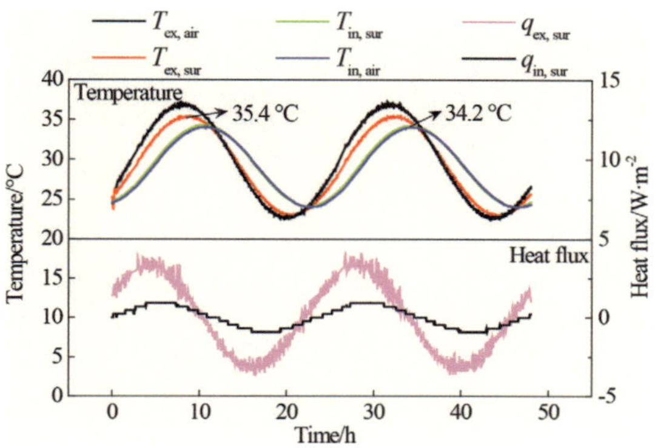

Figure 3.
Temperature under harmonic temperature fluctuation.

exterior and interior air temperature, the maximum temperature difference is 2.8° C, and the lag time is 2.5 h. The results prove that the insulation wallboard with poor thermal storage capacity is insufficient for the stability of indoor thermal environment.

Figure 4 shows the temperature and feat flux curves of the PCM-based lightweight wallboard (r_A = 65%) under harmonic temperature changing process. Due to the weak thermal inertia of insulation materials, the temperature changing of interior air is faster than the interior surface temperature changing of the PCM-based lightweight wallboard. It can be seen from the heat flux curves that the heat flux is delayed as much as 14.4 h. It is corresponding accurately between the largest temperature difference and the heat flux peak.

The temperature of the exterior/interior surfaces under the harmonic mode is shown in **Figure 5**. The temperature of interior surface and air is more stable in cell 1 (interior surface arrangement of PCM layer) than in cell 2 (exterior surface arrangement of PCM layer). The results show that the interior surface arrangement of PCM layer can reduce temperature amplitude in interior air to 32.1%. It proves that the PCM layer should be arranged on interior surface to improve thermal performance of PCM-based envelope.

Figure 6 illustrates the heat flux curves in PCM-based lightweight wallboards surfaces during the harmonic process. Because the PCMs are exposed to exterior air

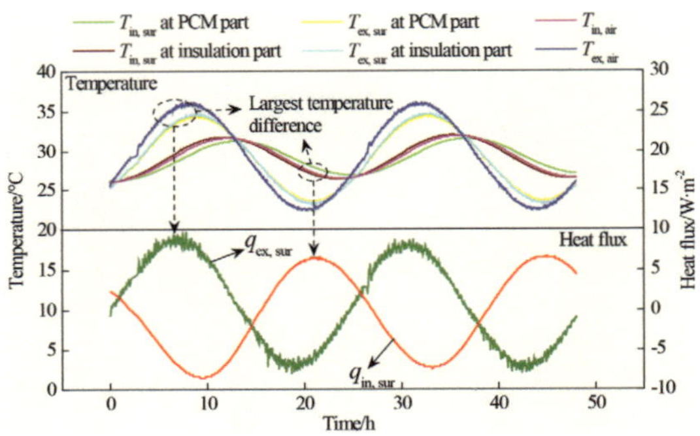

Figure 4.
Temperature and heat flux distribution of PCM-based lightweight wallboard during harmonic temperature process (r_A = 65%).

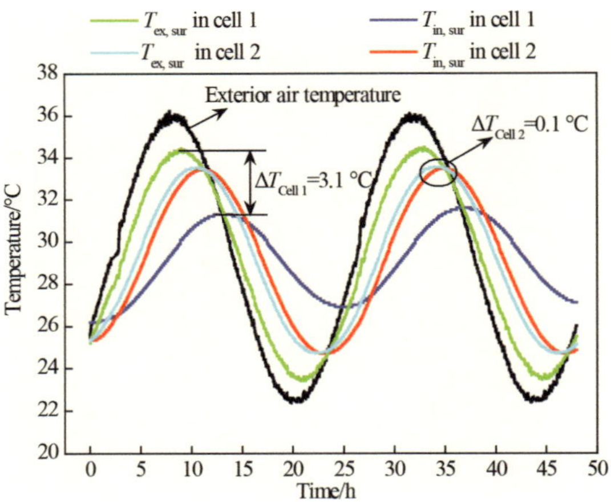

Figure 5.
Surface temperature curves during the harmonic process.

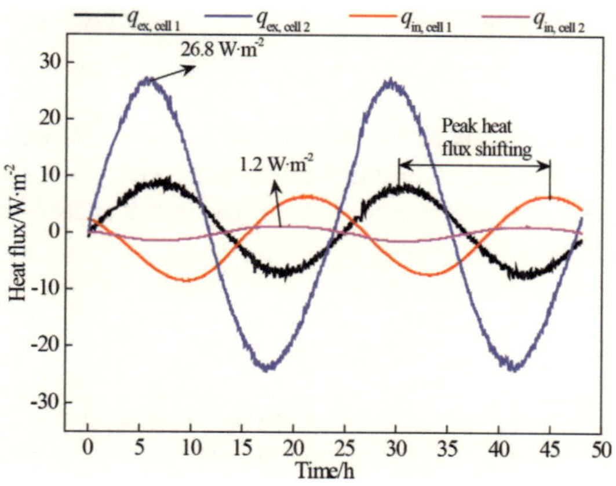

Figure 6.
Heat flux curves in PCM-based lightweight wallboards surfaces during the harmonic process.

in cell 2, the heat flux amplitude of exterior surface is larger than in cell 1. Meanwhile, because most of the incoming heat is absorbed by the PCM directly, the heat flux fluctuation on interior surface of cell 2 is less.

It can be proved by the experimental results that PCM layer with interior surface arrangement could achieve excellent performance to adjust indoor thermal environment. Although the PCM layer with exterior surface arrangement could absorb more heat from ambient environment, the PCMs cannot directly regulate the indoor thermal environment. By absorbing indoor excess heat, the PCM layer on interior surface can stabilize the indoor thermal environment. Therefore, it is a suitable approach to arrange the PCM layer on the interior surface of building envelope in the thermal design.

2.2 Thermal design method

As a traditional thermal design method of building envelope, harmonic response method can obtain the temperature fluctuation attenuation and delay time of normal envelope such as concrete and brick structures [11]. The method is widely accepted by engineers [12]. A simple thermal design method named HR-EC (harmonic response method and equivalent specific heat capacity principle) is proposed based on harmonic response method and equivalent specific heat capacity principle. The PCM-based envelope is different from normal envelope. During the phase change process, the latent heat will cause changing apparent specific heat. Therefore, to conveniently calculate the impact of latent heat, the calculation method adopts the equivalent specific heat capacity principle to simplify latent heat [13]. The temperature amplitude and delay time of interior surface of PCM-based lightweight wallboard are calculated and verified by comparing with the experimental results.

The HR-EC thermal design method is based on the harmonic response method, which is developed based on heat transfer analysis through building walls using Fourier transforms. Harmonic response method is a simple calculation method to obtain temperature damping and time delay of the walls without the analysis of heat transfer process. Therefore, it is a useful approach in the PCMs envelope thermal design.

Damping factor (v_0) of outdoor air temperature to wall interior surface temperature can be calculated as [12]

$$v_0 = 0.9e^{\frac{\Sigma D}{\sqrt{2}}} \cdot \frac{S_1 + \alpha_i}{S_1 + Y_{1,e}} \cdot \frac{S_2 + Y_{1,e}}{S_2 + Y_{2,e}} \cdot \cdots \frac{S_n + Y_{n\text{-}1,e}}{S_n + Y_{n,e}} \cdot \frac{\alpha_e + Y_{n,e}}{\alpha_e} \tag{2}$$

where ΣD is thermal inertia of building wall, S_n is coefficient of heat accumulation of wall materials, $Y_{n,e}$ is coefficient of heat accumulation of the outer surface of material layer, α_i is interior surface coefficient of heat transfer, and α_e is exterior surface coefficient of heat transfer.

D is thermal inertia of wall material layer. It can be defined as

$$D = R \cdot S \tag{3}$$

where R is thermal resistance of wall material layer and S is coefficient of heat accumulation of wall material. Thermal inertia represents the performance of material layers to resist the effects of fluctuating temperature, which is reflected by the temperature fluctuations of back of the material layer. ΣD can be obtained as

$$\Sigma D = D_1 + D_2 + \cdots + D_n \tag{4}$$

where D_1, D_2...D_n are the thermal inertia of each material layer of the wall.

S_n and Y_n are expressed as follows [12]:

$$S_n = \sqrt{\frac{2\pi\lambda c\rho}{Z}} \tag{5}$$

$$Y_n = \frac{R_n S_{n^2} + Y_{n\text{-}1}}{1 + R_n Y_{n\text{-}1}} \tag{6}$$

where λ is thermal conductivity, c is specific heat capacity, ρ is wall materials density, and Z is period of temperature fluctuation.

Phase delay (Φ_0) between the maximum temperature of outdoor air and the maximum temperature of wall interior surface is calculated by [12]

$$\Phi_0 = 40.5\sum D + \arctan\frac{Y_{ef}}{Y_{ef} + \alpha_e\sqrt{2}} - \arctan\frac{\alpha_i}{\alpha_i + Y_{if}\sqrt{2}} \tag{7}$$

where Y_{ef} is coefficient of heat accumulation of wall outer surface and Y_{if} is coefficient of heat accumulation of wall inner surface. Because the temperature period is 24 h in building thermal design, the delay time (ξ_0) can be obtained as [12]

$$\xi_0 = \frac{Z}{360}\Phi_0 = \frac{1}{15}\Phi_0 \tag{8}$$

Building envelope is always affected by double-side thermal effects including indoor and outdoor temperature changes. Therefore, the damping factor (v_{if}) and time delay (Φ_{if}) between indoor air and interior surface temperature of building envelope should also be obtained by the following Equations [12]:

$$v_{if} = 0.95\frac{\alpha_i + Y_{if}}{\alpha_i} \tag{9}$$

$$\Phi_{if} = \arctan\frac{Y_{if}}{Y_{if} + \alpha_i\sqrt{2}} \tag{10}$$

$$\xi_{\text{if}} = \frac{Z}{360}\Phi_{\text{if}} = \frac{1}{15}\Phi_{\text{if}} \tag{11}$$

It is necessary to determine the calculation parameters. The equivalent specific heat capacity principle is adopted to convert the latent heat to a constant value. The equivalent specific heat capacity (c_{p}) is defined as [13].

$$cp = \begin{cases} c_{\text{p}}^{\text{L}} & (T > TL) \\ c_{\text{p}}^{\text{M}} + c_{\text{p}}^{*} & (TL \geq T \geq TS) \\ c_{\text{p}}^{\text{S}} & (TS > T) \end{cases} \tag{12}$$

where c_{p}^{L} is the specific heat capacity in liquid phase, c_{p}^{S} is the specific heat capacity in solid phase, and c_{p}^{M} is the average value of c_{p}^{L} and c_{p}^{S}; c_{p}^{*} can be calculated as [13]

$$c_{\text{p}}^{*} = \frac{\Delta h}{\Delta T_{\text{R}}} \tag{13}$$

where ΔT_{R} is the temperature range during the complete phase change process and Δh is the latent heat of phase change process. Δh is obtained as

$$\Delta h = \Delta H \cdot ratio \tag{14}$$

where ΔH is the latent heat of the PCM and *ratio* is the phase change ratio of PCMs during phase change process. In the phase change process, the thermal conductivity of the PCM is also replaced as equivalent thermal conductivity [13]:

$$\lambda_{\text{p}} = \begin{cases} \lambda_{\text{p}}^{\text{L}} & (T > TL) \\ \lambda_{\text{p}}^{\text{S}} + \dfrac{\lambda_{\text{p}}^{\text{L}} - \lambda_{\text{p}}^{\text{S}}}{\Delta T_{\text{R}}} & (TL \geq T \geq TS) \\ \lambda_{\text{p}}^{\text{S}} & (TS > T) \end{cases} \tag{15}$$

where $\lambda_{\text{p}}^{\text{S}}$ is the thermal conductivity in solid phase and $\lambda_{\text{p}}^{\text{L}}$ is the thermal conductivity in liquid phase.

The results are listed in **Table 1**. The calculated results are close to the experimental results with 7.7% relative error. The results of interior surface temperature delay time have larger relative error with 33.2%. As an estimated method used in early design, the method is an easy way to predict the trend of temperature amplitude and delay time.

For the calculation of temperature damping factor and delay time by HR-EC method, the input parameters of wall materials are necessary to be determined, including thermal conductivity (λ), density (ρ), specific heat capacity (c), and material layer thickness (d). In addition, the equivalent specific heat capacity (c_{p}) is also a significant parameter for the calculation of PCM-based envelope thermal

	Calculated results	Experimental results	Relative error (%)
Temperature amplitude	4.06°C	4.40°C	7.7
Temperature delay time	3.38 h	5.06 h	33.2

Table 1.
Calculated and experimental results of PCM-based lightweight wallboard.

performance. It is critical to determine the phase change ratio in phase change process. In this part, it is determined based on experimental results. To provide references for PCM-based envelope thermal design in buildings, the phase change ratio of PCM-based envelope in different climate conditions could be obtained by series of further experiments.

3. How to deal with issues on climatic and seasonal suitability of the technology

The purpose of the present work is to investigate the seasonal and climatic suitability for application of PCM-based envelope coupled with night ventilation (NV) strategy in naturally ventilated office buildings. For suitability analysis, the adaptive thermal comfort theory is applied and the adaptive comfort model in standard ASHRAE-55 is adopted [14].

3.1 Numerical simulation details

Dynamic thermal simulations are performed using EnergyPlus. All simulations are carried out from April 1 to October 31, which covers the whole cooling season. The weather data is accessed based on the standard CSWD weather files provided by EnergyPlus [15]. The Conduction Finite Difference (CondFD) solution algorithm is adopted to depict the thermal process of phase change layer. The Fully Implicit Order is selected as the difference scheme for CondFD [16]. According to the suggestion of Tabares-Valesco et al. [17], time step for CondFD in EnergyPlus is set to 3 min, and node space is set to 3. The other components of the envelope are modeled with the default Conduction Transfer Functions algorithm. The heat conduction between the ground and the floor is calculated by Ground Heat Transfer: Slab module. TARP and DOE-2 are adopted for the interior and exterior surface convective heat transfer algorithms, respectively [18]. To perform NV strategy, the Zone Ventilation: Wind and Stack Open Area module is employed [19].

Three-dimensional model and standard floor plan of the office building are depicted in **Figure** 7. The four-storey building is 3.6 m storey height, 31.9 m length, and 15.8m width. The total floor area is 2016 m^2. The size of each window is 1.8m height and 1.5 m width. The window-wall ratio is 16.3% of the north and south walls and 5.1% of the remaining walls, which satisfies local standard [20].

According to Ref. [21], it has been verified that phase change temperature (*PCT*) has great influence on thermal performance of PCMs in different seasons.

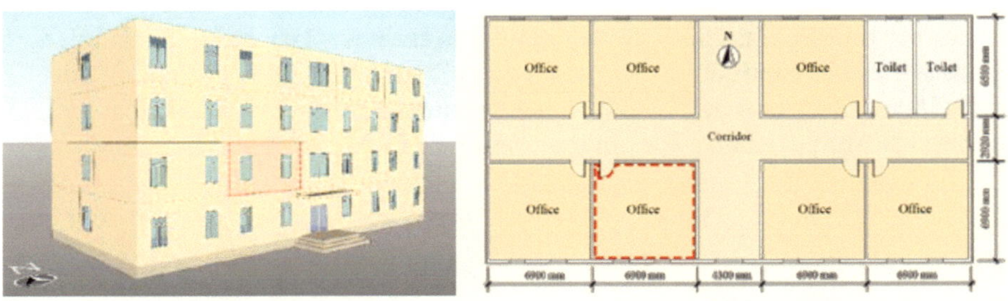

(a) (b)

Figure 7.
(a) The 3D model and (b) the standard floor plan of the building for EnergyPlus simulation.

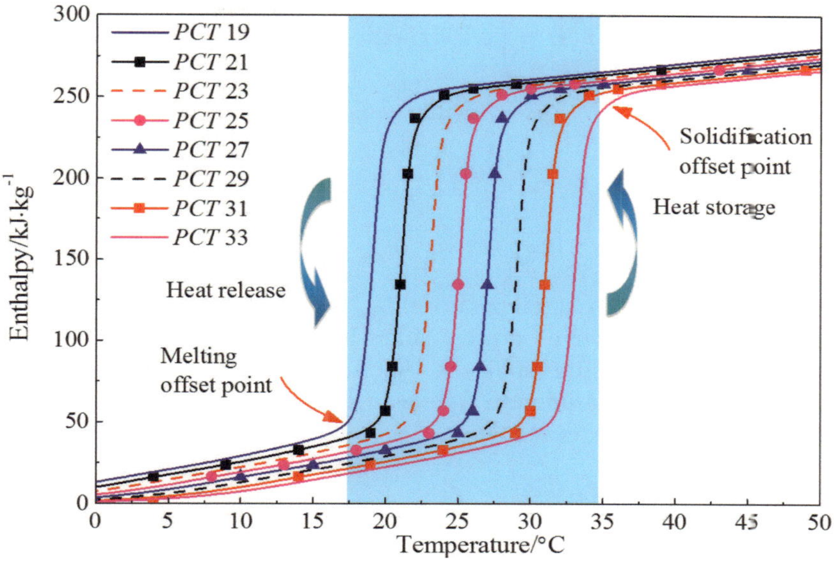

Figure 8.
H-T curves of eight different PCMs.

Therefore, to examine the climatic and seasonal suitability of *PCT*, Bio-PCMs with eight different *PCT* are adopted. The enthalpy-temperature graph of Bio-PCMs is depicted in **Figure 8** [22]. The latent heat of Bio-PCMs is 219 kJ·kg^{-1}. The density is 860 kg·m^{-3} and the specific heat is 1.97 (kJ·kg^{-1}·^{o}C^{-1}). The thermal conductivity is 0.2 (W·m^{-1}·^{o}C^{-1}). Bio-PCMs are installed on the inner side of building envelope (external walls, internal walls and ceilings) in Ref. [23].

3.2 Evaluation index

With the adaptive comfort theory, Evola et al. [24] proposed an index called Intensity of Thermal Discomfort (*ITD*), to evaluate the indoor thermal environment. It is defined as the time integral, over the occupancy period of the positive difference between the current operative temperature and the upper limit of thermal comfort range (see **Figure 9**). This index reveals well the duration and the extent of discomfort thermal sensation perceived by the occupants within a long period. Therefore, it is employed to represent the thermal performance of PCMs coupled with NV strategy. The calculation is expressed as follows:

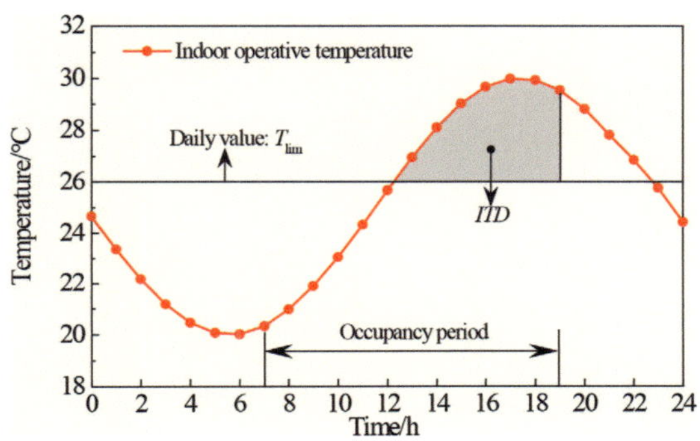

Figure 9.
Definition of ITD.

$$ITD = \int_{P} \Delta T^{+}(\tau) \cdot \mathrm{d}\tau$$
$$\text{where } \Delta T^{+}(\tau) = \begin{cases} T_{op}(\tau) - T_{lim} & \text{if } T_{op}(\tau) \geq T_{lim} \\ 0 & \text{if } T_{op}(\tau) < T_{lim} \end{cases} \quad (16)$$

$$T_{lim} = 0.31 \cdot T_{rm} + 21.3 \quad (17)$$

where T_{op} is the operative temperature, T_{lim} is the upper threshold of 80% acceptability limit of Adaptive Comfort Model in Standard ASHRAE-55 [14], and T_{rm} is the mean monthly outdoor air dry-bulb temperature, which is the arithmetic average of the mean daily minimum and mean daily maximum outdoor air dry-bulb temperature for the month in sequence.

3.3 Climatic and seasonal suitability analysis

A series of parametric studies over the selected eight Bio-PCMs are carried out, to investigate the suitability of *PCT* in transition and hot seasons. The optimum *PCT* for transition season and hot season is determined according to the minimum *ITD*, which are listed in **Table 2**. It can be found that in some cities, the optimal *PCT* for transition season does not match with the climate conditions in hot season, vice versa. Therefore, it is quite necessary to select different *PCT* based on the outdoor climatic characteristics of transition and hot seasons, respectively.

Based on the optimum *PCT*, the effect of PCMs coupled with NV strategy on reducing *ITD* in transition and hot seasons is compared with NV strategy and PCMs strategy. **Figure 10** summarizes the *ITD* in both transition and hot seasons with different technologies in all cities. All strategies (with PCMs coupled with NV strategy, with NV strategy, and with PCMs strategy) can reduce *ITD* compared with reference group (without cooling strategy). The *ITD* of transition season is shown in **Figure 10(a)**. For these cities in hot summer and warm winter zone, the effects with PCMs alone in reducing *ITD* are inferior to NV strategy. It is evident for Nanning and Hechi, where the *ITD* with PCMs is much higher than that with NV. When NV is introduced on the basis of PCMs strategy, *ITD* is reduced from 2231 to

Cities	Optimum *PCT*/°C	
	Transition season	Hot season
Turpan	27	33
Nanning	29	29
Hechi	27	29
Chongqing	27	29
Xi'an	25	27
Chengdu	25	27
Urumqi	23	25
Altay	23	25
Guiyang	23	25
Kunming	23	23

Table 2.
Optimum PCT for transition season and hot season.

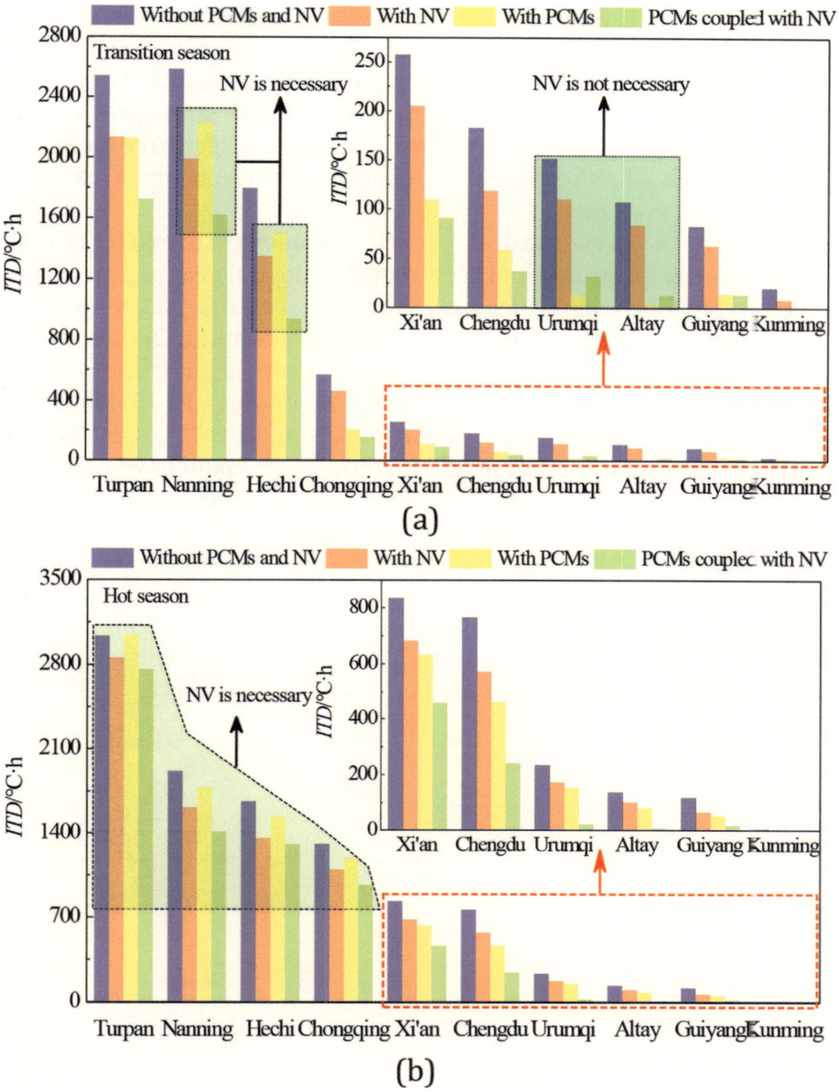

Figure 10.
(a) Transition season and (b) hot season of the influence of different passive cooling technologies on ITD.

1621°C·h for Nanning and from 1503 to 937°C·h for Hechi. It is due to NV strategy working well to introduce outdoor cool air into the room and to exclude the heat released from PCMs to outdoor during nighttime. Accordingly, PCMs can be fully solidified at night. Therefore, *ITD* can be effectively decreased during the occupancy period using PCMs coupled with NV. It further indicates the coupling use of NV strategy, and PCMs strategy is necessary for transition season under this climate condition. On the contrary, PCMs coupled with NV strategy is inferior to PCMs strategy for severe cold zones. It is clear that *ITD* is increased when PCMs coupled with NV strategy is adopted in Urumqi and Altay compared to PCM strategy. Such results indicate that PCM strategy is more suitable to severe cold zone in transition season rather than PCMs coupled with NV strategy. For other cities, PCMs coupled with NV strategy is more suitable in transition season in reducing *ITD* compared to NV strategy and PCM strategy.

From **Figure 10(b)**, PCMs coupled with NV is the most effective strategy to reduce *ITD* in hot season for all cities. However, the advantages of PCMs coupled with NV strategy than NV strategy in Turpan, Nanning, Hechi, and Chongqing are greatly reduced compared to transition season due to the high temperature and small diurnal temperature difference in hot season. Additionally, with PCM strategy alone, *ITD* in these cities is much higher than that with NV strategy or PCMs coupled with NV strategy. It indicates that NV strategy is necessary for PCM strategy to obtain excellent performance. The use of PCMs coupled with NV strategy is critical to hot season.

A performance-based inverse design method is proposed to extract typical outdoor air dry-bulb temperature from typical meteorological year, which is favorable for application of PCMs coupled with NV. The typical air dry-bulb temperature curve (T_{out}) is obtained as **Figure 11** shown (Nanning, China). Outdoor air dry-bulb temperature fluctuation range is much larger than the phase transition temperature range (27–31°C). Meanwhile, the maximum value T_{max} (33°C) is 2°C higher than the upper limit of *PCT*, and the minimum value T_{min} (25.5°C) is 1.5°C lower than the lower limit of *PCT*. In other words, PCMs could have an opportunity to solidify at low temperature phase and to melt at high temperature phase. Additionally, the average value T_{ave} (28.6°C) is quite close to *PCT* 29°C, and the diurnal temperature difference (ΔT) is greater than 7.5°C. These conditions are beneficial to the utilization of latent heat of PCMs.

Figure 12 is the comparison of T_{out} and outdoor air dry-bulb temperature in a typical week (1–7 June). T_{out} agrees well with outdoor air dry-bulb temperature curve. It means the approach can be used to determine T_{out} curve.

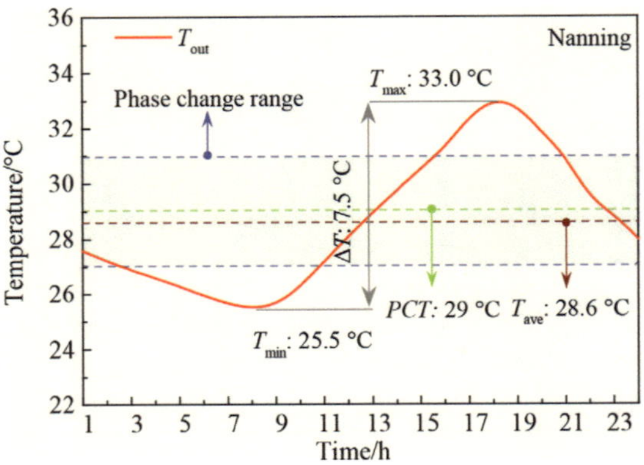

Figure 11.
Characteristics of the typical outdoor air dry-bulb temperature curve.

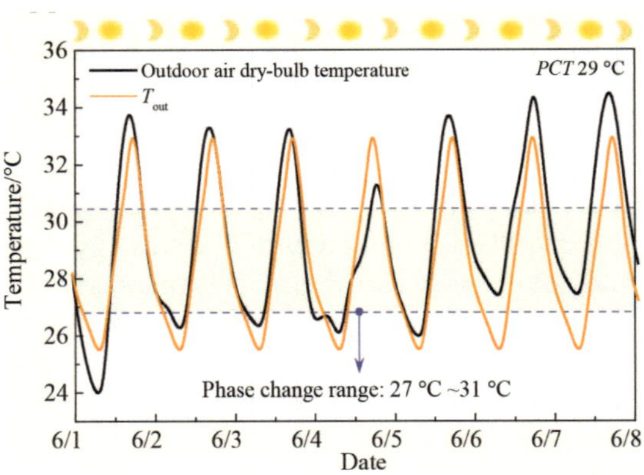

Figure 12.
Comparison between T_{out} and outdoor air dry-bulb temperature in a typical week.

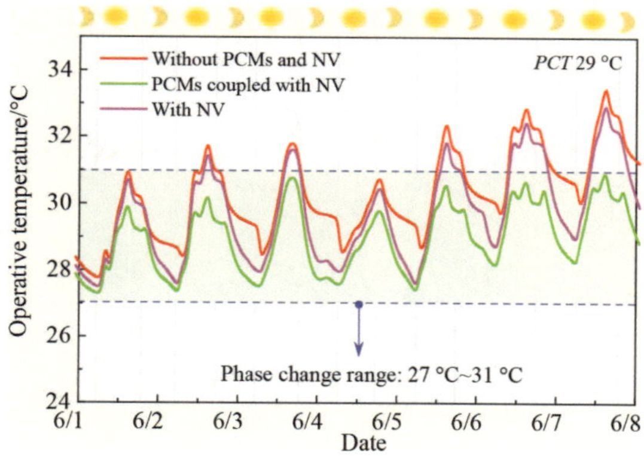

Figure 13.
Indoor operative temperature in the typical week: PCMs coupled with NV vs. NV.

In order to verify the reliability of the method, indoor operative temperature using PCMs coupled with NV and NV in the typical week is compared in **Figure 13**. PCMs coupled with NV appear as a remarkable advantage in reducing indoor operative temperature than NV. Indoor operative temperature reduction is up to 1°C, even to 2°C in 6–8 June. It can infer that outdoor air dry-bulb temperature in the typical week is beneficial to PCMs coupled with NV. It further indicates that the method is reliable. Therefore, it can be used to estimate the relationship between PCMs coupled with NV strategy and local outdoor climatic characteristics.

4. How to improve thermal performance of phase change materials applied in building envelope

Taking PCMs into building envelope significantly reduces building energy consumption and improves indoor thermal environment. The main categories of PCMs applied in building envelope are organic PCMs and inorganic PCMs. They both have solid–liquid phase flow properties, and the phase transition takes place at a certain temperature range. Paraffin is the most commonly used organic because it has chemically stable, high latent of fusion, and regular degradation in thermal properties after phase change cycles. However, paraffin has low thermal conductivity, flammable, and slow oxidation when exposed to oxygen. These problems present challenges in container design [25]. Salt hydrates as an important group of inorganic PCMs have extensively investigated in building envelope. The most serious limitation of the salt hydrates is phase segregation and supercooling compared with paraffin. Another problem is salt hydrates would cause corrosion in building materials and metal containers. These problems limit their applications in building envelope [26].

The integration of PCMs and building envelope can be divided into three types: direct incorporation and immersion, macro-encapsulation, and micro-encapsulation (see **Figure 14**). Using PCMs in building envelope, one must keep in mind that liquid phase causes leaking to the surface out of the carrier materials. Encapsulation serves as barrier between PCM and surrounding environment. It provides long-term durability and structural requirements. Micro-encapsulation of PCMs provides faster charging and discharging rates because of the smaller distance for heat transfer compared to macro-encapsulation. However, the lower encapsulation rate of micro-encapsulation greatly reduced the energy storage and increases the cost [27].

Figure 14.
Macro-encapsulation of PCMs applied in building envelope [28, 29].

4.1 Heat transfer enhancement at material level

Among the different combinations of PCMs and building envelope, macro-encapsulation (building component with PCMs) is considered as a promising option in building energy conservation. The advantages of the technique embodied at various packaging shapes and sizes includes avoiding leakage of material, providing sufficient thermal storage capacity, and reducing the adverse effects on structure, whereas unreasonable application and poor thermal conductivity would cause the regulating ability of indoor thermal environment untimely and incompletely. The building applications require fast absorption and release of latent heat. In addition, small temperature fluctuation is not beneficial to the utilization of PCMs in buildings. Hence, heat transfer enhancement of PCMs is significant for building envelope. The heat transfer performance of PCMs building component can enhance at material level and component level [30]. At material level, carbon-based materials are one of the most common additives due to high thermal conductivity, stable chemical performance, and low density. However, uniformly dispersion of additives in PCMs is the most important factor to improve thermal conductivity. Moreover, aspect ratio of additives is also crucial for thermal conductivity enhancement such as carbon fiber (CF) and carbon nanotubes (CNTs).

At material level, CNTs is selected as a heat transfer medium for the composite PCM in experiment. A group of the composite PCM is prepared with CNTs of 0.5, 1, 1.5, and 2 wt%. The thermal conductivity and latent heat are determined and compared. The experimental results show that thermal conductivity of the composite PCM with 2 wt% CNTs increased by 17 and 21% compared to pure paraffin in liquid and solid phase, respectively.

To evaluate the heat storage and release capacity of pure paraffin and the composite PCM, the infrared thermograph images of two samples are recorded in **Tables 3** and **4**. It can be found that the heat transfer rate of the composite PCM is greatly improved both in heat release and heat storage process. Especially, the average temperature of the composite PCM at different time is 2°C higher than that of pure paraffin in heat storage process. Moreover, the temperature distribution of the composite PCM is more uniform than that of pure paraffin in heat storage process.

4.2 Heat transfer enhancement at component level

At component level, metal fins are widely used to increase the heat transfer areas and relatively easy to fabricate for building component with PCMs [31]. The main

Time		0 min	10 min	20 min	30 min	Legend
Heat release	Pure paraffin					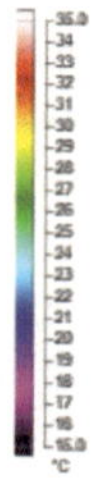
	T_{ave}	26°C	24.7°C	23.2°C	23.1 °C	
	Composite PCM					
	T_{ave}	26°C	24.1°C	23.0°C	22.8 °C	

Table 3.
Heat release comparison of pure paraffin and the composite PCM.

Time		0 min	5 min	10 min	15 min	Legend
Heat storage	Pure paraffin	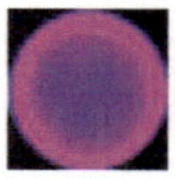	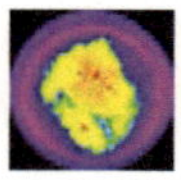	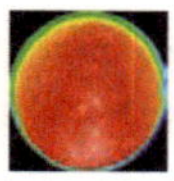	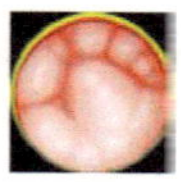	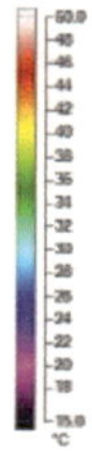
	T_{ave}	22.5°C	36.9°C	45.2°C	48.3°C	
	Composite PCM	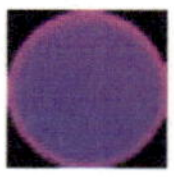	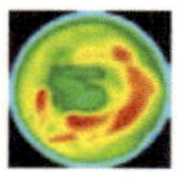			
	T_{ave}	22.2°C	38.8°C	47.6°C	49.3°C	

Table 4.
Heat storage comparison of pure paraffin and the composite PCM.

Time	45 min	60 min	75 min	90 min	Legend
Non-optimized component (top and bottom)					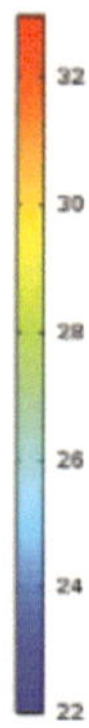

Optimized component (top and bottom)		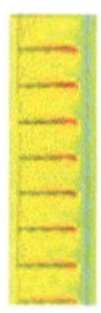	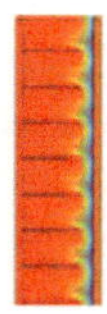	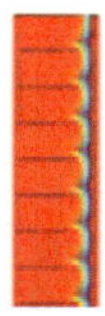	
		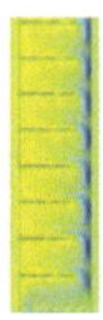			

Table 5.
Comparisons of optimized/nonoptimized component with PCMs.

influence factors on thermal performance of PCMs component are fins number and geometry. The current research works mainly focused on the effect of fins number or geometry separately. However, these two factors should be taken into consideration. To enhance heat transfer rate, the balance between fins number and geometry should be further investigated.

In the field of PCMs heat transfer, numerical method is widely employed by architects. To obtain the appropriate geometric of fins, the building component with PCMs is optimized by COMSOL Multiphysics. The results show that PCMs complete melting time in the building component and the temperature difference of heat transfer surfaces could be used to evaluate the fins effect. The optimized building component presents the best heat transfer performance with 8 mm length, 0.2 mm width, and 5 mm spacing of fins. The temperature comparison of building component with PCMs before and after optimization is shown in **Table 5**. It can be seen that the heat transfer performance of the optimized component is better than non-optimized component during heat storage process. The temperature difference between top and bottom in the non-optimized component is increasing over the time. It directly gives the fact that appropriate geometry of fins makes the heat transfer much more uniform and reduce the adverse effect of natural convection in heat storage process.

5. What is the application mode?

PCMs coupled with NV are an effective passive cooling strategy for office buildings. A kind of inorganic PCMs encapsulated in PVC panel developed by a Chinese manufacturer is applied in building envelope as a wall application in a naturally ventilated office building. According to differential scanning calorimeter (DSC) test [32], the latent heat of the PCM is 122.3 $kJ \cdot kg^{-1}$ in melting and 116.9 $kJ \cdot kg^{-1}$ in solidification. The phase change temperature range is 25–27°C. The thickness of PVC panel is 20 mm, and the length and width are 520 mm and 200 mm. The PCM panel is installed at the interior surface of an external wall and an internal wall of the office building through modular installation as shown in **Figure 15**. The PCM panel is directly connected to the wall using aluminum strips and nuts by the modular installation.

For north wall, the PCM panel is directly installed on the interior surface (see **Figure 16(a)**). The PCM panel exposed to indoor air is beneficial to convection heat transfer between indoor air and the PCM panel. At nighttime, the outdoor cool air is introduced into indoor space through the opened window. When the cold air flows through the PCM-based wall, the coldness is stored by PCMs. For south wall, the

Figure 15.
(a) Isometric view of the naturally ventilated office building and (b) diagram of installation position of PCM panel.

DOI: http://dx.doi.org/10.5772/intechopen.85012

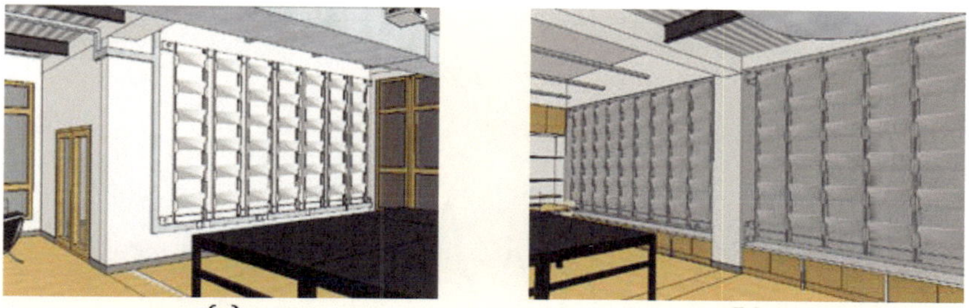

(a) (b)

Figure 16.
(a) Design sketch, PCM panel exposed to indoor air; and (b) design sketch, PCM panel decorated with perforated aluminum plate.

installation of the PCM panel is the same as that of the north wall. However, considering that there is a constructional column in the middle of the south wall protruding into the interior space, the exterior of the PCM panel is decorated with perforated aluminum plate. These tiny holes in the aluminum plate can promote the convection heat transfer between indoor air and the PCM panel (see **Figure 16(b)**).

Actual installation effect of the PCM panel is shown in **Figure 17**. To examine the application effect of PCMs coupled with NV strategy in hot season, full-scale field thermal environment test is conducted in a typical week of hot season. Indoor air temperature, surface temperature, and heat flux of the wall with the PCM panel are carefully investigated. To obtain the cooling potential of PCMs coupled with NV, the scenario of the wall without the PCM panel is modeled using EnergyPlus with the actual weather data during the test period. The accuracy and reliability of the method is validated by comparing the simulated and experimental results with maximum relative error of 2.95% and average relative error of 0.96%.

The indoor thermal environment before and after the application of PCMs coupled with NV is compared with experiment and simulation results. It can be observed from **Figure 18(a)** that the interior surface heat flux of the wall with the PCM panel is significantly greater than that without the PCM panel. It demonstrates that PCMs absorb and release a lot of heat from indoor air during the melting or solidification process. Therefore, the interior surface peak temperature of the wall with the PCM panel is greatly reduced with respect to wall without the PCM panel as shown in **Figure 18(b)**. Accordingly, the indoor air peak temperature is greatly reduced using the PCM panel. The maximum reduction of indoor air temperature is up to 1.1°C (see **Figure 18(c)**).

(a) (b)

Figure 17.
Actual installation effect of the PCM panel: (a) exposed to indoor air and (b) decorated with perforated aluminum plate.

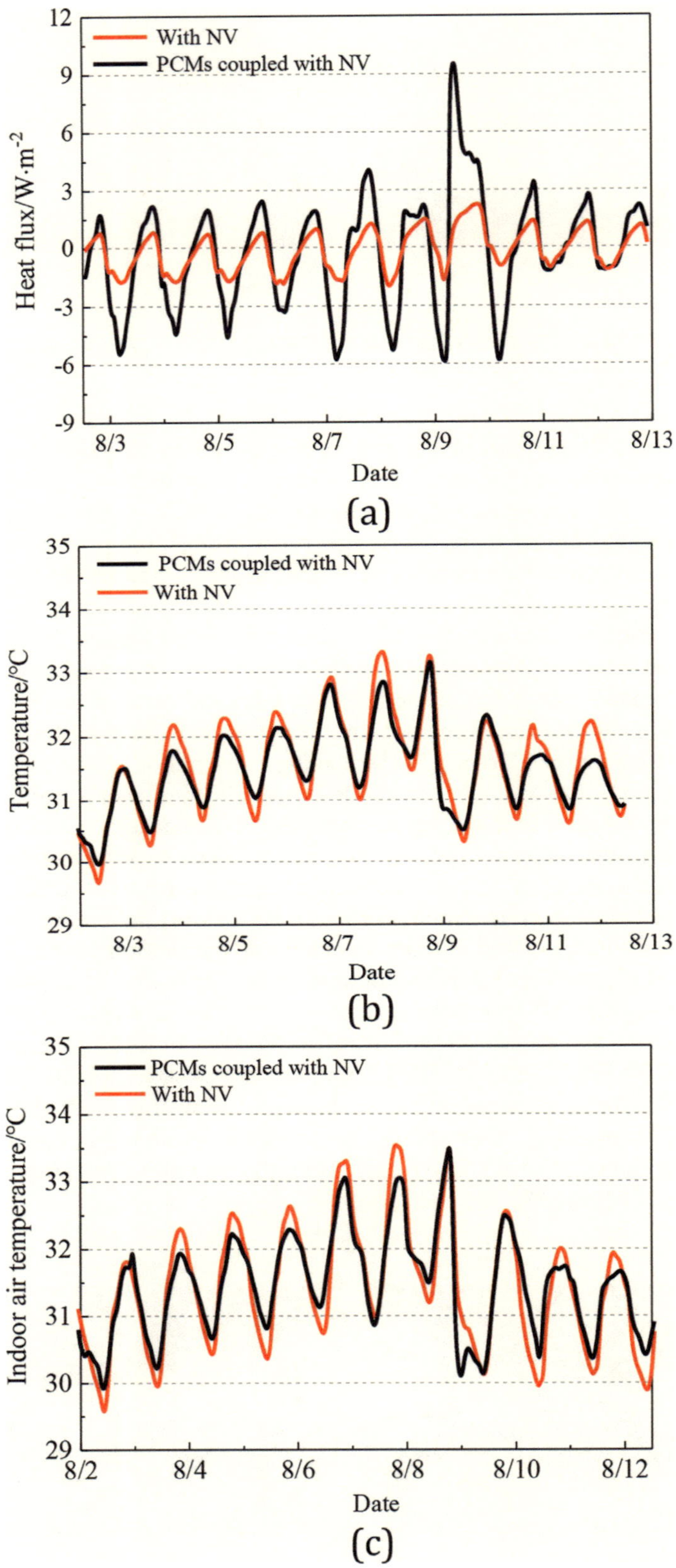

Figure 18.
(a) Interior surface heat flux of north wall, (b) interior surface temperature of north wall, and (c) indoor air temperature.

6. Conclusions

1. Comparative experimental investigations of thermal performance of PCM-based lightweight wallboards are conducted in the artificial controlled condition with harmonic temperature changing. The thermal performance of PCM-based lightweight wallboards is carefully analyzed. Two kinds of controlled conditions are compared: different PCM layout areas and different PCM layer arrangements are adopted. Moreover, a simple thermal design method is proposed and verified with the artificial controlled experimental results.

2. A typical naturally ventilated office building has been modeled with EnergyPlus, and several simulations have been carried out based on 10 representative cities of Western China. The climatic and seasonal suitability for application of PCMs coupled with NV strategy is investigated from three aspects: suitability of *PCT*, comparisons of three different cooling strategies, and application advantage. Moreover, a performance-based inverse design method is proposed to determine the typical climate characteristics.

3. Based on two methods of thermal transfer enhancement of PCMs applied in building envelope, PCMs and building component with PCMs are optimized in material and component level, respectively. The optimum proportion of CNTs in the composite PCM is obtained through experiment. Appropriate geometry of fins is optimized by numerical simulation. Results show that two optimization methods can promote the heat transfer performance of PCMs and improve the utilization of PCMs.

4. A kind of PCM panel is applied as wall application in a naturally ventilated office building. Full-scale field indoor thermal environment test is conducted in a typical week of hot season. The thermal performance of the wall with PCM panel is superior to the wall without the PCM panel. The heat storage capacity of the wall can be improved by 40%, and the indoor air peak temperature can be reduced by 1.1°C in the test period of hot season.

Acknowledgements

This work was supported by “the 13th Five-Year” National Science and Technology Major Project of China (No. 2018YFC0704500), Natural Science Foundation of China (No. 51838011 and No. 51808429), Shaanxi Province Key Research and Development Plan (No. 2017ZDXM-SF-076), China Postdoctoral Science Foundation (No. 2018 T111026), and Natural Science Foundation of Shaanxi Province (No. 2017JQ5005).

Conflict of interest

We declare that there is no conflict of interest exits in the chapter, as well as the chapter is approved by all authors for publication.

Nomenclature

A_{ALL}	Area of the wallboard, m^2
A_{PCM}	Area of PCM, m^2

c	Specific heat capacity of materials, kJ·kg^{-1}·°C^{-1}
c_p	Equivalent specific heat capacity, kJ·kg^{-1}·°C^{-1}
D	Thermal inertia of wall material layer, dimensionless
d	Thickness of wall materials, m
ITD	Intensity of thermal discomfort, °C·h
PCT	Phase change temperature, °C
$q_{ex,sur}$	Heat flux of exterior surface, W·m^{-2}
$q_{in,sur}$	Heat flux of interior surface, W·m^{-2}
R	Thermal resistance of wall material layer, m^{2}·°C·W^{-1}
r_A	Ratio of PCM part area to wallboard area, dimensionless
$ratio$	Phase change ratio of PCMs during phase change process, %
S	Coefficient of heat accumulation of wall materials, W·m^{-2}·°C^{-1}
T_{ave}	Average value of typical outdoor air dry-bulb temperature, °C
$T_{ex.air}$	Exterior air temperature, °C
$T_{ex.sur}$	Temperature of the exterior surface, °C
$T_{in,air}$	Interior air temperature, °C
$T_{in.sur}$	Temperature of the interior surface, °C
T_{lim}	Upper threshold of 80% acceptability limit of Adaptive Comfort Model in Standard ASHRAE-55, °C
T_{max}	Maximum value of typical outdoor air dry-bulb temperature, °C
T_{min}	Minimum value of typical outdoor air dry-bulb temperature, °C
T_{op}	Operative temperature, °C
T_{out}	Typical outdoor air dry-bulb temperature, °C
T_{rm}	Mean monthly outdoor air dry-bulb temperature, °C
v_{if}	Damping factor between indoor air and interior surface temperature of building envelope, dimensionless
v_0	Damping factor of outdoor air temperature to wall interior surface temperature, dimensionless
Y_{ef}	Coefficient of heat accumulation of wall outer surface, W·m^{-2}·°C^{-1}
Y_{if}	Coefficient of heat accumulation of wall inner surface, W·m^{-2}·°C^{-1}
$Y_{n,e}$	Coefficient of heat accumulation of the outer surface of material layer, W·m^{-2}·°C^{-1}
Z	Period of temperature fluctuation, h
Greek letters	
α_e	Exterior surface coefficient of heat transfer, W·m^{-2}·°C^{-1}
α_i	Interior surface coefficient of heat transfer, W·m^{-2}·°C^{-1}
ΔH	Latent heat of the PCM, kJ·kg^{-1}
Δh	Latent heat of phase change process, kJ·kg^{-1}
ΔT	Diurnal temperature difference, °C
ΔT_R	Temperature range during the complete phase change process, °C
ΔT^{+}	Difference between operative temperature and comfort temperature, °C
λ	Thermal conductivity of materials, W·m^{-1}·°C^{-1}
λ_p	Thermal conductivity of PCMs, W·m^{-1}·°C^{-1}
ξ_{if}	Delay time between the highest indoor temperature and the inner surface temperature of wall, h
ξ_0	Delay time between the highest outdoor temperature and the inner surface temperature of wall, h
ρ	Density of materials, kg·m^{-3}
Φ_{if}	Phase difference between the highest indoor temperature and the inner surface temperature of wall, deg

Φ_0	Phase difference between the highest outdoor temperature and the inner surface temperature of wall, deg
Abbreviations	
CF	Carbon fiber
CondFD	Conduction finite difference
CNTs	Carbon nanotubes
DSC	Differential scanning calorimeter
EPS	Expandable polystyrene
HR-EC	Harmonic response method and equivalent specific heat capacity principle
NV	Night ventilation
PCMs	Phase change materials
TSV	Thermal storage and ventilation
TSVL	Thermal storage and ventilation laboratory

Author details

Liu Yang[1,2]*, Yan Liu[1,2], Yuhao Qiao[1,2], Jiang Liu[1,2] and Mengyuan Wang[1,2]

1 State Key Laboratory of Green Building in Western China, Xi'an University of Architecture and Technology, Xi'an, Shaanxi, P.R. China

2 School of Architecture, Xi'an University of Architecture and Technology, Xi'an, Shaanxi, P.R. China

*Address all correspondence to: yangliu@xauat.edu.cn

References

[1] Souayfane F, Fardoun F, Biwole PH. Phase change materials (PCM) for cooling applications in buildings: A review. Energy and Buildings. 2016;**129**: 396-431. DOI: 10.1016/j.enbuild.2016.04.006

[2] Liu Y, Yang L, Zheng WX, Liu T, Zhang XR, Liu JP. A novel building energy efficiency evaluation index: Establishment of calculation model and application. Energy Conversion and Management. 2018;**166**:522-533. DOI: 10.1016/j.enconman.2018.03.090

[3] Yang L, Lam JC, Tsang CL. Energy performance of building envelopes in different climate zones in China. Applied Energy. 2008;**85**(9):800-817. DOI: 10.1016/j.apenergy.2007.11.002

[4] Navarro L, Gracia AD, Niall D, Castell A, Browne M, Mccormack SJ. Thermal energy storage in building integrated thermal systems: A review. Part 2. Integration as passive system. Renewable Energy. 2016;**85**(2): 1334-1356. DOI: 10.1016/j.renene.2015.06.064

[5] Liu Y, Yang L, Hou LQ, Li SY, Yang J, Wang QW. A porous building approach for modelling flow and heat transfer around and inside an isolated building on night ventilation and thermal mass. Energy. 2017;**141**: 1914-1927. DOI: 10.1016/j.energy.2017.11.137

[6] Lam JC, Yang L, Liu JP. Development of passive design zones in China using bioclimatic approach. Energy Conversion and Management. 2006;**47**(6):746-762. DOI: 10.1016/j.enconman.2005.05.025

[7] Liu J, Liu Y, Yang L, Hou LQ, Wang MY, Liu JP. Annual energy saving potential for integrated application of phase change envelopes and HVAC in Western China. Procedia Engineering. 2017;**205**:2470-2477. DOI: 10.1016/j.proeng.2017.09.975

[8] Li DHW, Yang L, Lam JC. Zero energy buildings and sustainable development implications—A review. Energy. 2013;**54**(6):1-10. DOI: 10.1016/j.energy.2013.01.070

[9] Zhang YP, Lin KP, Zhang Q, Di HF. Ideal thermophysical properties for freecooling (or heating) buildings with constant thermal physical property material. Energy and Buildings. 2006;**38**(10):1164-1170. DOI: 10.1016/j.enbuild.2006.01.008

[10] Kuznik F, David D, Johannes K, Roux JJ. A review on phase change materials integrated in building walls. Renewable and Sustainable Energy Reviews. 2011;**15**(1):379-391. DOI: 10.1016/j.rser.2010.08.019

[11] Zhou JL, Zhang GQ, Lin YL, Li YG. Coupling of thermal mass and natural ventilation in buildings. Energy and Buildings. 2008;**40**(6):979-986. DOI: 10.1016/j.enbuild.2007.08.001

[12] Chinese National Standard. Thermal Design Code for Civil Building (GB 50176–1993). Beijing: Ministry of Housing and Urban-Rural Development of the People Republic of China; 1993. [in Chinese]

[13] Chou HM, Chen CR, Nguyen VL. A new design of metal-sheet cool roof using PCM. Energy and Buildings. 2013;**57**(1):42-50. DOI: 10.1016/j.enbuild.2012.10.030

[14] ASHRAE Standard 55-2010. Thermal Environmental Conditions for Human Occupancy. Atlanta: American Society of Heating, Refrigerating and Air-conditioning Engineers; 2010

[15] EnergyPlus: Weather Data [Internet]. 2019. Available from: https://

www.energyplus.net/weather [Access Time: 06-10-2019]

[16] Al-Saadi SN, Zhai ZQ. Modeling phase change materials embedded in building enclosure: A review. Renewable and Sustainable Energy Reviews. 2013;**21**:659-673. DOI: 10.1016/j.rser.2013.01.024

[17] Tabares-Velasco PC, Christensen C, Bianchi M. Verification and validation of EnergyPlus phase change material model for opaque wall assemblies. Building and Environment. 2012;**54**: 186-196. DOI: 10.1016/j.buildenv.2012.02.019

[18] EnergyPlus: Engineering Reference [Internet]. 2019. Available from: https://www.energyplus.net/documentation [Access Time: 06-01-2019]

[19] Jamil H, Alam M, Sanjayan J, Wilson J. Investigation of PCM as retrofitting option to enhance occupant thermal comfort in a modern residential building. Energy and Buildings. 2016; **133**:217-229. DOI: 10.1016/j.enbuild.2016.09.064

[20] Chinese National Standard. Design Standard for Energy Efficiency of Public Buildings (GB 50189-2015). Beijing: Ministry of Housing and Urban-Rural Development of the People Republic of China; 2015. [in Chinese]

[21] Saffari M, Gracia AD, Fernández C, Cabeza LF. Simulation-based optimization of PCM melting temperature to improve the energy performance in buildings. Applied Energy. 2017;**202**:420-434. DOI: 10.1016/j.apenergy.2017.05.107

[22] Muruganantham K. Application of phase change material in buildings: Field data vs EnergyPlus simulation [thesis]. Arizona: Arizona State University; 2010

[23] Cui YP, Xie JC, Liu JP, Wang JP, Chen SQ. A review on phase change material application in building. Advances in Mechanical Engineering. 2017;**9**(6):1-15. DOI: 168781401770082

[24] Evola G, Marletta L, Sicurella F. A methodology for investigating the effectiveness of PCM wallboards for summer thermal comfort in buildings. Building and Environment. 2013;**59**: 517-527. DOI: 10.1016/j.buildenv.2012.09.021

[25] Sharma SD, Sagara K. Latent heat storage materials and systems: A review. International Journal of Green Energy. 2005;**2**(1):1-56. DOI: 10.1081/GE-200051299

[26] Baetens R, Jelle BP, Gustavsen A. Phase change materials for building applications: A state-of-the-art review. Energy and Buildings. 2010;**42**(9): 1361-1368. DOI: 10.1016/j.enbuild.2010.03.026

[27] Akeiber H, Nejat P, Majid MZA, Wahida MA, Jomehzadeh F, Famileh IZ, et al. A review on phase change material (PCM) for sustainable passive cooling in building envelopes. Renewable and Sustainable Energy Reviews. 2016;**60**: 1470-1497. DOI: 10.1016/j.rser.2016.03.036

[28] Silva T, Vicente R, Soares N, Ferreira V. Experimental testing and numerical modelling of masonry wall solution with PCM incorporation: A passive construction solution. Energy and Buildings. 2012;**49**:235-245. DOI: 10.1016/j.enbuild.2012.02.010

[29] Castell A, Martorell I, Medrano M, Pérez G, Cabeza LF. Experimental study of using PCM in brick constructive solutions for passive cooling. Energy and Buildings. 2010;**42**(4):534-540. DOI: 10.1016/j.enbuild.2009.10.022

[30] Liu LK, Su D, Tang YJ, Fang GY. Thermal conductivity enhancement of phase change materials for thermal energy storage: A review. Renewable

and Sustainable Energy Reviews. 2011; **15**(1):24-46. DOI: 10.1016/j.rser.2010.08.007

[31] Ma T, Yang HX, Zhang YP, Lu L, Wang X. Using phase change materials in photovoltaic systems for thermal regulation and electrical efficiency improvement: A review and outlook. Renewable and Sustainable Energy Reviews. 2015;**43**:1273-1284. DOI: 10.1016/j.rser.2014.12.003

[32] Yang L, Qiao YH, Liu Y, Zhang XR, Zhang C, Liu JP. A kind of PCMs-based lightweight wallboards: Artificial controlled condition experiments and thermal design method investigation. Building and Environment. 2018;**144**: 194-207. DOI: 10.1016/j.buildenv.2018.08.020

Chapter 3

Advances in Passive Cooling Design: An Integrated Design Approach

Ahmed A.Y. Freewan

Abstract

Incorporating passive cooling devices within building design requires analysis of device variables and actions to improve cooling performance, maximize efficiency, and integrate with building elements. Improving devices performance requires understanding the relation of devices to design stages, building elements, and working mechanism, and actions performed by devices to enhance cooling process and effectiveness. Therefore, designers could integrate passive devices as intrinsic design elements. The current research introduces *SARS* as an innovative classification of passive devices based on cooling actions that are performed by a device like *storing*, *avoidance*, *removal* or *slowing (SARS)*. All actions, devices, and variables were discussed and analyzed to help integrate them within design stages: analysis, designing, and performance. Understanding actions will help maximize the performance of the devices, combine two or more devices together, and integrate the devices' deign in design process. Combining more devices together to perform more than one function will move passive design to a new level to become as whole building design approach and to be a core design element.

Keywords: passive cooling, storing, avoidance, removal, slowing, design matrix, shading, ventilation

1. Introduction

The integration of passive systems in the architectural design process requires many considerations on all levels of design stages. The aim of this integration is to achieve and provide high-efficiency thermal comfort or natural lighting. Passive system performances depend mostly on natural and environmental elements like the sun, wind, earth, and water. It is, therefore, significant to study and analyze how passive systems interact with natural elements and their relationship to a building site. Passive cooling systems, thereafter, need to be integrated within the design process, as their performance requirements are affected by orientation, height, materials, form, and characteristics of many architectural elements.

The main two innovative ideas of this chapter are, first, to provide analysis of passive devices based on one or more of the cooling actions (store, avoid, remove, and slow (SARS)) and, second, to develop a design matrix based on SARS actions and analysis of examples of integrated passive systems.

2. Passive cooling systems

Passive cooling systems were the main design driver in low energy architecture and vernacular architecture especially in hot climate regions. Buildings were designed and built to adapt to local environmental conditions and to use natural elements to provide occupants with the required thermal comfort around the year. However, the discovery of fossil fuels and the development of new construction materials the building industry became less respect to the surrounding environment and less dependent on passive techniques. Buildings become heavily dependent on energy to provide the indoor environment with the required thermal comfort. The use of fossil fuels on a large spreader scale resulted on many environmental and health problems that associated with the greenhouse emissions. These problems shifted architects' awareness to environmental and climatic variables to be key drivers of the building design to provide the required thermal comfort and reduce energy consumptions.

Passive cooling or heating approaches play a major role in bringing architecture closer to the original green, environmental, and vernacular architecture by using surrounding environmental elements such as solar radiation and natural ventilation. Passive cooling is an approach that focuses on providing thermal comfort by controlling heat gains and heat dissipation without involving mechanical or electrical devices. The performance quality of this approach depends totally on the interaction of the building's design and devices with the surrounding environmental factors, such as sun rays, ambient air temperature, wind, and humidity, to achieve energy balance for occupants. Therefore, conducting a thorough analysis of a building's local climatic conditions is essential for any passive cooling approach to successfully fulfill its purpose and to maximize a specific action of SARS.

Heat gain sources include internal and external sources. The internal heat gains are produced from human activities, artificial lights, equipment, and appliances used by the occupants, while the external heat gains result from the interaction of the building with the outdoor environment. Heat gain or loss has four forms: *first*, heat gains caused by solar radiation passing through opaque envelope materials and heating the interior spaces with the greenhouse effect, *second*, heat gain caused by direct sun rays transmitted through windows and transparent surfaces into the interior spaces, *third*, heat gains caused by conduction between the building envelope and the surrounding environment, and, *fourth*, heat gains through convection caused by air infiltration and ventilation exchange between the outdoor and indoor environment as seen in **Figure 1**.

Lechner [1] presented many simple passive devices used for cooling in hot climates like simple courtyard, wind tower, thermal massing, windows, arcade, shading devices, and solar chimney. In addition the book discussed the ventilation principles and pattern. Many studies investigated in different approaches and methods the potential of passive cooling techniques in saving energy and cooling the indoor space in different climates. Nunes and Oliveira Panão [2] suggested a method to calculate the monthly cooling energy needs in zones where passive cooling systems are installed and applied on an office ventilated by passive devices like earth cooling and solar chimney. Kachkouch et al. [3] investigated three passive techniques in the real conditions in hot semiarid climate. The study tested the effect of three main techniques are, shading ceiling color and insulation on heat flux. The study concluded that these techniques helped to reduce heat flux with best results of white painted. Prieto et al. [4] showed how important it is to apply the passive strategies in early design stages and use active equipment after that if necessary. They studied, using simulation software, the effectiveness of selected passive cooling strategies like glazing, shading, color, and heat sink in commercial

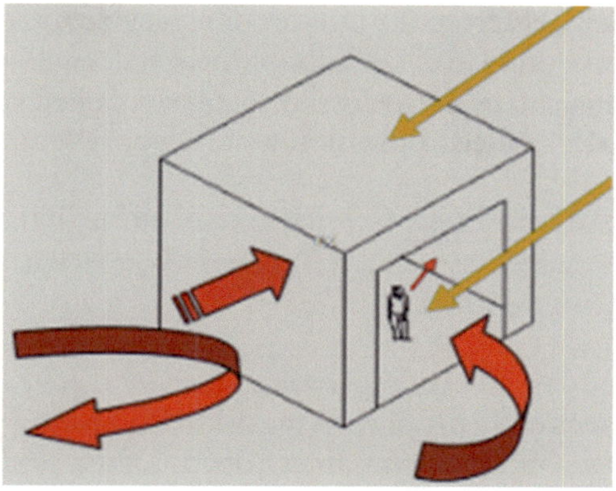

Figure 1.
The interaction of the building with the environmental elements (author).

buildings in warm climates. The study concluded that the efficiency of the passive strategies conditioned to both the harshness of a given climate and design of different building parameters. Tejero-González et al. [5] reviewed many passive design techniques and parameters affecting applicability of such techniques. The study investigated how the climatic parameters are needed to be studied thoroughly to design and select passive cooling/heating techniques. Panchabikesan et al. [6] studied many passive design techniques like evaporative cooling, nocturnal radiative cooling, and phase change material (PCM) in different climatic conditions in India to reduce energy consumption. The study showed that these techniques best result in hot climates. Oropeza-Perez and Østergaard [7] review many passive and active cooling methods that could be used in residential buildings. They studied firstly many technologies in term of heat balance, secondly they scientifically analyzed the results, and finally they focused on feasibility and economic value of the findings. The study developed a decision-making program to find out the most suitable cooling method in dwelling design. Passive cooling focuses on controlling heat gain or heat loss in a building in order to reduce energy consumption and in order to create the indoor thermal comfort [8].

3. Passive cooling actions

Understanding the sources of heat gains that affect thermal comfort in the building is essential for deciding the type of actions to be taken to avoid as much heat gains as possible, to slow the heating process to remove the uncontrolled gained heat, or to store cold air or elements. The four passive cooling actions include the following:

1. *Storing* of cold mass or air within building envelope. This action is defined by keeping cold air or mass away from direct heat gains to provide spaces with cold air or cool down the air before entering the interior spaces like courtyards, basements, earth spaces, and thermal masses.

2. *Avoidance* of direct external solar radiation heat gain. This action is conducted by applying design considerations and devices in the building. Avoidance could be applied by using shading windows and glazed areas, using landscape, designing of self-shading forms, and considering colors and reflectivity of external surfaces.

3. *Removal* of gained heat from the interior or exterior sources. This action is required to remove portion of undesirable heat that could not be avoided or slowed. The action can be performed through controlled ventilation, by using wind towers, earth tunnels, and windows to support ventilation requirements.

4. *Slowing* heat transfer from the external climate through the building envelope. This action is conducted by using techniques like efficient insulation and double glazing window units.

The four actions of passive cooling, *storing*, *avoidance*, *slowing*, and *removal*, are discussed in this section in detail with regard to the cooling principles and the designing process, focusing on the variables affecting their performances. Each device or principal will be discussed based on the following categories:

- The required actions of implementation from designers in each design process, in stages of early stage (analysis), middle stage (design), and final stage (performance)
- The building variables related to the devices and principles
- The required conditions of using the device
- Using case studies and examples of integration of devices and actions in the buildings' design

Some passive systems have direct impact on the architecture design, like the design of an opened atrium or a courtyard, and some have less direct influence, like the type of material and use of louvers or shading systems. Therefore, the level of integration between these systems and the architectural elements varies among the different types of systems and actions. However, it is important to understand the requirements of the chosen passive systems and to make the right decisions of when and where to integrate them within the designing process.

3.1 Storing devices

Storing refers to keep cold air or low temperature object away from direct heat sources to be used to cool down interior spaces. Throughout architectural history, and particularly in hot climate regions, devices were developed for this purpose as storage design elements, like courtyards, earth spaces, basements combined with wind towers, thermal masses, and sunken courtyards. For the purpose of this research, courtyards are discussed as one of the most efficient devices in storing cold objects or air.

3.1.1 Courtyard

Courtyard is an opened space surrounded by rooms with openings on the conjoint wall between the rooms and the courtyard space, to allow air exchange daylight, and view. The courtyard as passive design device works as a modifier of the microclimate and acts as a heat sink and cold air storage. Buildings with internal courtyards are characterized as a suitable solution for cooling in hot climate regions to provide inner spaces with cold air and daylight.

The working mechanism of the courtyard depends on the cycle of day and night, which results on a continuous change of air temperature and the difference in air temperature between the inside and outside of the courtyard (**Figure 2**). Therefore the

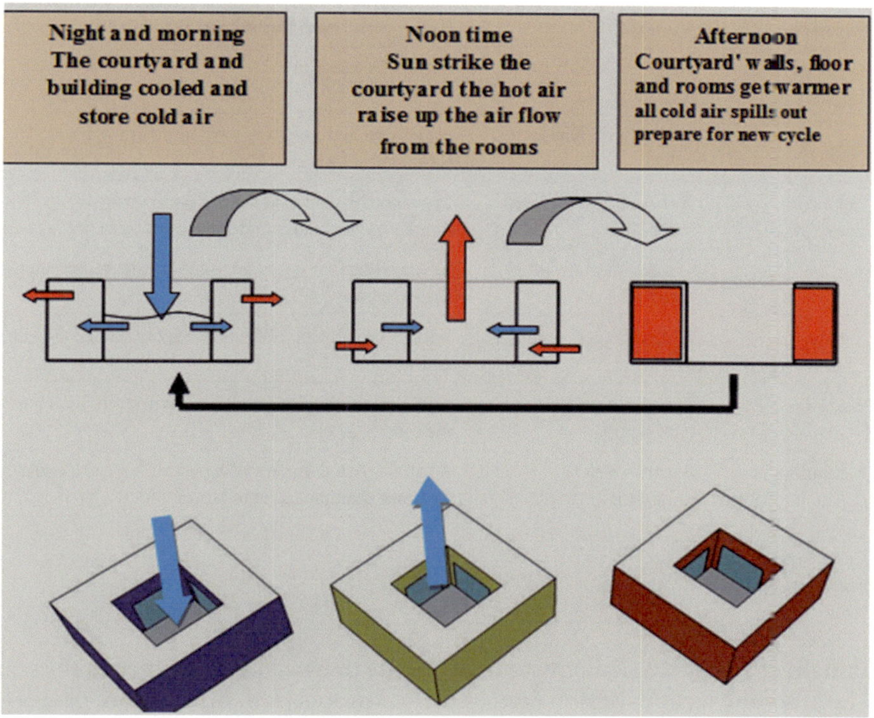

Figure 2.
The three periods in the working mechanism of a courtyard (author).

performance quality of the courtyard house depends on the heat exchange processes between the indoor spaces and the courtyard and then between the courtyard space and the external open spaces. It consists of three main periods; during the first one, the cool night air sinks into the courtyard and flows to the surrounding rooms, and therefore the spaces and surfaces are cooled until noon time. During this period, the courtyard works as storage of cold air and cold air exchange with surrounding rooms. The second period starts at noon, when the sun strikes the floor of the courtyard directly and the temperature of air inside the court's space starts increasing gradually, causing the hot air to move up, and consequently, air is drawn from surrounding rooms to the courtyard space through the openings, resulting in cooling the surrounding rooms. The last period starts when the courtyard and the surrounding rooms get warmer, and all cool air leaks out, which prepares the system for a new cycle in the next day.

3.1.1.1 Courtyard design variables to maximize storing effect

The passive cooling performance of courtyards depends on three types of elements: the courtyard elements like walls, floor, and landscape greeneries; building elements like windows and built space characteristics; and finally site elements like location, climate, and orientation as summarized in **Table 1**. The cooling performance of courtyards depends on providing the enclosed spaces protection from direct solar radiation and controlled airflow by studying the orientation of the courtyard with regard to the solar path and providing trees for shades and designing with consideration of prevailing winds. **Table 1** summarizes the different variables affecting the cooling performance of a courtyard that designers need to take in consideration in the designing process:

Optimizing the passive cooling performance of courtyards in hot regions depends mainly on how to design an efficient storage space by studying shading patterns, geometrical shape, and thermal mass material.

Elements	Variables	How to maximize storing action in hot climate
Walls, floor	Height-to-width or to area ratio	High H/W ratio [11]
	Material, thickness, color	High thermal mass stores low temperature, heat absorbance, and reflection
Geometrical shape	Inside stepped or tilted walls [27]	Increases the shaded areas and improves courtyard's thermal performance and storing action in hot regions
	Outside stepped or tilted [27]	Increases surface exposure to sun and therefore reduces the shaded areas
Landscape	Greenery type, height	Define the patterns of shade or expose shaded area on floor and walls
Windows	Location, window/ wall ratio	As the openings allow air to be drawn from the rooms to the courtyard
Orientation	North–south or west–east	A balance based on prevailing winds and solar angles, to optimize shading and ventilation

Table 1.
Courtyard design variables (author).

Zamani et al. [9] studied how to improve the thermal performance of the courtyard by studying various design factors such as proportion, orientation, geometry, opening characteristics, and material. In addition they studied more variables like shading devices, vegetation, and water pools and their impact on heat mitigation.

Sadafi et al. [10] explained how using internal courtyards in terraced houses in tropical regions improves the natural ventilation and thermal comfort. Meir et al. [11] investigated how two semi-enclosed attached courtyards will affect the microclimate in the enclosed courtyards and the attached built volume. Muhaisen [12] showed that courtyards' shading performance depends on the form's properties, location, latitude, and available climatic conditions. Berkovic et al. [13] studied the effect of courtyard design variables, like orientation, horizontal shadings, galleries, and trees, on the thermal comfort of the courtyard's surrounding functions. Muhaisen and Gadi [14] investigated how courtyards' proportions and surface colors considerably influence thermal comfort of the surrounding spaces. Al-dawoud and Clark [15] investigated how different design parameters of the courtyard affect the thermal comfort in spaces surrounding the courtyard. They approved that courtyards are more energy efficient in hot-dry and hot-humid climates in comparison to cold climates.

The design of courtyard with other devices and elements should be made to enhance storing effect and ventilation process, which include the opening, thermal mass, landscape, and building form. Elements that could be integrated should enhance storing action of the courtyard like thermal mass, wall geometry, and landscape. Moreover, to enhance heat exchange between the courtyard and surrounding spaces, other devices could be integrated with the courtyards like wind tower, solar chimney, basement, and opening design and bearing in mind the design requirements in early stages of design process. Therefore, design decisions will have direct impacts on the building's form, orientation, area, zoning, function distribution, and relations with the outdoor and site design.

3.2 Avoidance

Avoidance, as a passive cooling action, refers to all the methods used to prevent and reduce the amounts of heat gains from direct solar radiation or wind. The key methods of avoidance include different shading devices, building's form, and landscape. Additional factors, including building's orientations and surfaces' colors and textures, can help to prevent gained heat from reaching inner spaces.

3.2.1 Shading devices

Cho et al. [16] presented an integrated approach for exterior shading device design analysis that included cooling energy performance and economic feasibility in high-rise residential buildings. The research investigated the effect of 48 exterior shading devices on the sunshading/daylighting performance. Palmero-Marrero and OLiveira [17] studied the effect of static louver shading devices on east, west, and south facades for various locations on the energy demands during cooling and heating seasons. The research concluded that the shading device reduced the total annual energy demands in buildings of countries with long dominant cooling seasons and high ambient temperatures and solar radiation.

Datta [18] studied the effect of external fixed horizontal louvers on the thermal performance in the buildings. The study was aimed for reducing the overall energy requirements for the entire year by maximizing the shading device system to reduce solar gains during summer and allow them during winter. The study used TRNSYS as a simulation to maximize the efficiency of the device, and different slat lengths and tilt angles were tested in four Italian cities. Yao [19] evaluated the effect of shading control strategies on the daylighting, visual comfort, and energy performance in buildings.

Designing buildings with the passive approach requires integration of many factors together in the process, such as orientation, shading devices, and building form in order to reduce energy consumption in the building as a whole as seen in **Table 2**. Largely glazed facades and large windows have been increasingly used in new buildings, allowing access to daylight, solar heat gains, and external views. The increase in glazed surfaces requires significant attention in building design, regarding the impact they have on cooling, heating, and lighting loads demands. Therefore, it is important to provide these buildings with a proper shading design that would provide interior spaces with thermal comfort by controlling solar heat gains and reducing glare while maintaining the initial purpose of large glazed surfaces to provide external views and sufficient daylighting.

Many researches were conducted to study the performance of shading devices in order to optimize their performance, save energy, and achieve the maximum thermal comfort. Datta [18] used computer simulation to study variables related to horizontal shading devices and their effect on the thermal performance in buildings in Italy. The study showed that shading devices could help save energy and

Elements	Variables	How to maximize avoidance actions
Types by	Horizontal louvers	Southern windows block high solar angles
	Vertical louvers	East and west windows block low solar angels
	Diagonal or eggcrate	Block low and high angles on east, south, and west directions
	Overhangs, canopy	The depth and height, considering solar noon in summer and winter
Device's variables	Space to depth	Depth-to-spacing ratio to balance between sunrays block and view out
	Material	To reflect or to absorb sunrays

Table 2.
Shading device variables (author).

improve the thermal performance of the buildings. Palmero-Marrero and OLiveira [17] proved that shading devices could improve thermal performance of buildings and save energy in many cities in different latitudes and climatic conditions.

3.2.1.1 Shading in existing buildings and new buildings

A study that has been conducted at Jordan University of Science and Technology to design shading devices showed that the process of designing shading devices for an existing building required reflections of various parameters besides thermal comfort. Many tools were used to monitor the performance of shading devices like patterns of use and users' behavior in the new setting. The study used real measurements, computer simulation, user's survey, and observation usage. The study showed that user's preferences like view out, natural lighting, illuminance levels, and thermal comfort were the most influential indicators in designing shading devices. Moreover, user's behavior and patterns of use in the office in question were monitored and proved that a well-designed shading device can improve thermal comfort, user satisfaction, and user behavior from energy consumption point of view. The study was conducted in two stages: in the first stage, temporary materials were used to study the integration

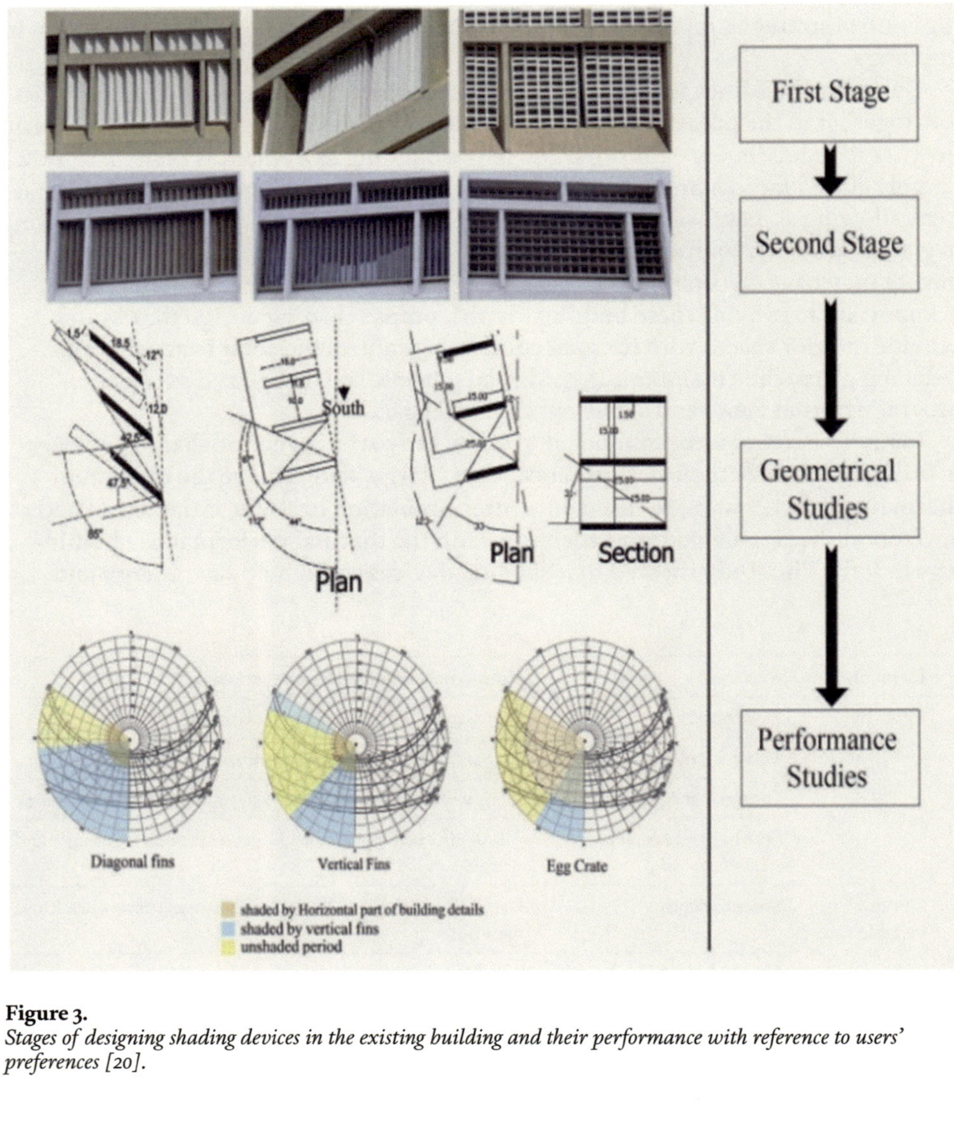

Figure 3.
Stages of designing shading devices in the existing building and their performance with reference to users' preferences [20].

of all variables as seen in **Figure 3**. In the second stage, permanent materials were used, and long-term monitoring was conducted in order to generalize the best design for shading devices in offices with large windows at the university campus [20].

3.2.1.2 Self-shading and building form

Hemsath and Alagheband Bandhosseini [21] stated that building form and orientation as early decisions in the design process could have a great impact on energy consumption, lighting, cooling, and heating load. Authors emphasized on the relation between the building's forms, shapes, daylight, and energy consumption in the early design phase, instead of using mechanical and artificial light. Many non-rectangular shapes had been evaluated in terms of self-shading and energy consumption like an L or U shape that can offer solar advantages. Zhou et al. [22] studied how optimal building design could enable harvesting of the maximal micro-wind power around low-rise residential buildings.

Zerefos et al. [23] compared polygonal and prismatic building envelopes to orthogonal building envelopes based on energy behavior and energy consumption in Mediterranean climates. The study showed that prismatic formed buildings gain lower solar than orthogonal forms and so consume less energy by an average of 7.88%. Moreover, Caruso et al. [24] used the mathematical theory of calculus of variation to find the best geometric form to minimize direct solar irradiation incident on the envelope. The paper also aimed at finding useful guidelines and rules for designers to follow during early decision-making stages to reduce the total amount of direct solar irradiation [24].

Azari et al. [25] showed that there are various architectural features of a building that could influence its indoor thermal comfort, daylight, and energy consumption, such as building shape, orientation, wall forms, window-to-wall area ratio, window size, glazing material, wall structure, and shading. They may increase solar gain and daylight duration during winter, which would be beneficial and could lead to overheating during summer. Yasa [26] studied the comfort conditions of different configurations of buildings like open courtyard configurations, closed courtyard configurations, and configurations of courtyards with apertures on the wall.

Designers can use building's forms as self-shading approach that shades the outside surface materials, windows, and glazed areas (**Table 3**). An example of a self-shading building is the library of the University of Nottingham, UK, which was designed with large glazing surfaces to utilize daylighting without causing glare as seen in **Figure 4**. The design was based on using self-shading form to protect inner spaces from direct sunrays.

Element's construct	Concepts	Variables	Notes
Building's elements	Exterior wall tilt angle	Winter and summer solar angle	Studying the maximum and minimum solar angles to determine the optimum tilt angle of the external walls for the building to act as self-shading form
Site's elements	Climate	Hot climates	Self-shading forms for buildings with large glazed surfaces and high requirements of daylighting
		Cold climates	Requires less self-shading forms and increase of the direct heat gains

Table 3.
Self-shading and building form variables.

Figure 4.
Library building at University of Nottingham, UK (author).

3.2.1.3 Landscape

Trees and landscape can improve the thermal environment and reduce the temperatures of interior spaces and surfaces in buildings. Monitoring tree effects on buildings showed that trees and landscapes do not only provide shade and reduce air temperature but also prevent buildings' materials from gaining and storing heat and radiating it back later as seen in **Table 4**. A study conducted in Jordan University of Science and Technology compared two identical glazed corridors of orientation and materials, but one of them with trees and landscape providing shade, and one without trees. The study showed that shaded building's surfaces with high trees recoded a lower temperature of 34°C than identical spaces with temperature of 41°C in summer day as seen in **Figure 5**.

Designing with avoidance systems requires analysis of the solar angles laterally with the dimensions and orientation of the building's site to identify the type of shading devices with integration and function and daylight strategies. Designer's decision of the building's form should consider self-shading building, orientation, elevation design, and relation to landscape to enhance avoidance actions. Performance evaluation of avoidance actions required integration of cooling decisions with all building elements, such as glazing area, orientation form and mass design, facade design, and opening.

3.3 Removal

Removing, as a passive cooling action, refers to the removal of undesirable gained heat in interior or exterior spaces in a building. Natural ventilation normally is the main strategy used to take unwanted heat out of the buildings. It depends mostly on pressure differences to circulate air between inner and outer spaces, allowing air to enter or escape from buildings. Devices like windows, openings,

Variables	Notes
Tree type	Evergreen or deciduous trees affect the periods and area of shading on buildings, walls, courtyards, and outdoor spaces in summer and winter
Height and horizontal spread	The pattern of shade provided by different types of trees
Distance from building	Location of shade on building, walls, walking area, and courtyards

Table 4.
Landscape variables (author).

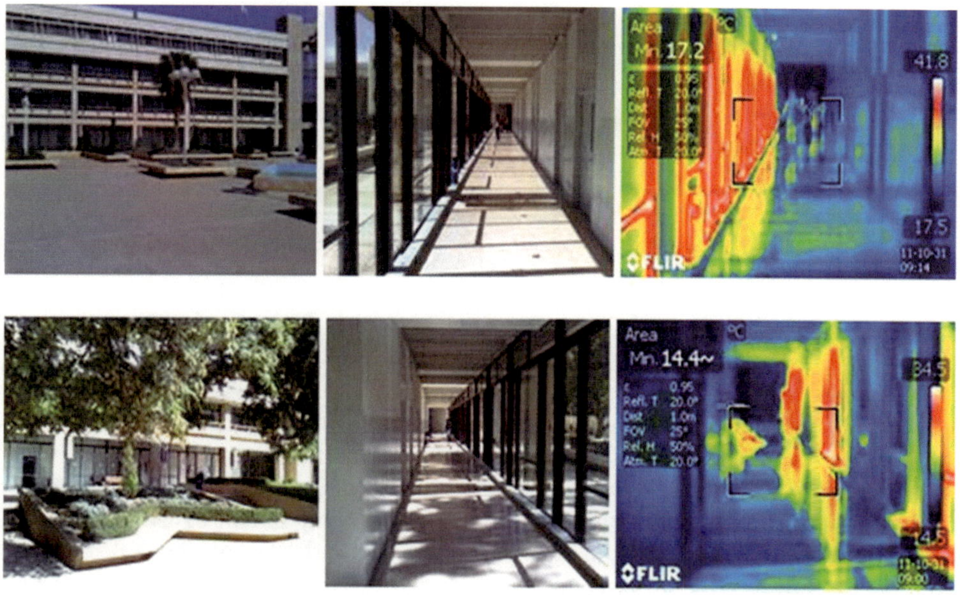

Figure 5.
The effects of trees on surface and air temperatures in identical spaces (author).

wind towers, sunspaces, earth tunnels, and roof openings are used to move air through a building.

Natural ventilation follows three main principles that designers should understand well in order to induce passive ventilation into any building's design, which includes stack ventilation, Bernoulli's effect, and the Venturi effect. These three principles uses air pressure differences due to height, air temperature, or wind speed, to pull air to or from buildings. Therefore, the main concept of passive ventilation design is achieved by maximizing the use of one or more of the three principles in order to induce natural ventilation.

3.3.1 Stack effect

Stack effect depends on temperature differences to circulate air, as hot air rises up and cool air sinks down (**Figure 6**). The design for ventilation that depends on stack principles is achieved by letting hot air rise up within spaces or specific devices and exhausting it from upper openings, which allows it to be replaced by cooler air from lower openings. The designer's role in the process is represented in designing air movement and its exhaustion and penetration, which includes the following methods:

1. Accelerating the rising of hot air by designing a long vertical space that crosses through the building section, like atriums, double-skin facades, solar chimneys, or wind towers

2. Designing inlet openings for cold air to enter the building from a well-planned and controlled cold space like shaded courtyards or urban spaces, basements, etc.

3. Increasing warm air to activate the stack effect and accelerate the ventilation process using devices like sunspaces, solar chimneys, and skylights in the building's design

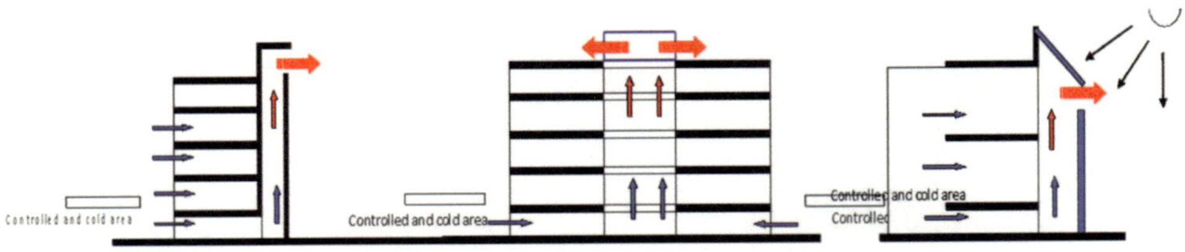

Figure 6.
Stack effect (author).

3.3.2 Bernoulli's effect

Bernoulli's effect depends on the reduction of air pressure associated with the wind speed. In buildings, designers use wind speed differences to circulate air inside or around the building **Figure 7**. Air movement around and above buildings creates positive and negative pressures, causing fresh air to be sucked through specific openings into the buildings at the same time allowing hot air to escape though designate openings and locations. The designer's role in the process is represented in planning and designing air movements with regard to the negative and positive air pressures zones, and it includes the following methods:

1. Designing the building's surroundings with the least possible obstructions to allow air flow around, creating the necessary positive and negative pressure zones.

2. Designing the building's form to go with the direction of the wind rather than obstructing it; this is to increase wind speed around the building and create positive and negative pressures.

3. Designing openings in the areas of positive and negative pressures with integration with the interior space distributions to maximize ventilation process.

3.3.3 Venturi effect

Venturi effect causes acceleration in air speed when it passes from a wide section area to a thinner section area, developing a negative pressure zone at the thinning points, which help suck the air from near spaces as seen in **Figure 8**. Designers can make use of this effect in building's designs by:

1. Designing urban settings, landscape, and group buildings with regard to the large scale to allow them to capture wind and increase its speed

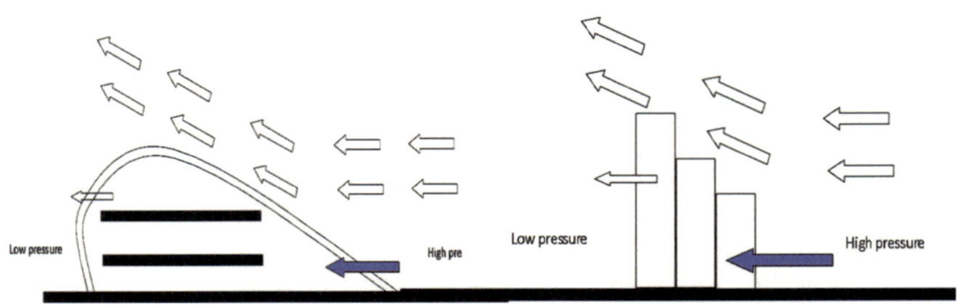

Figure 7.
Bernoulli's effect (author).

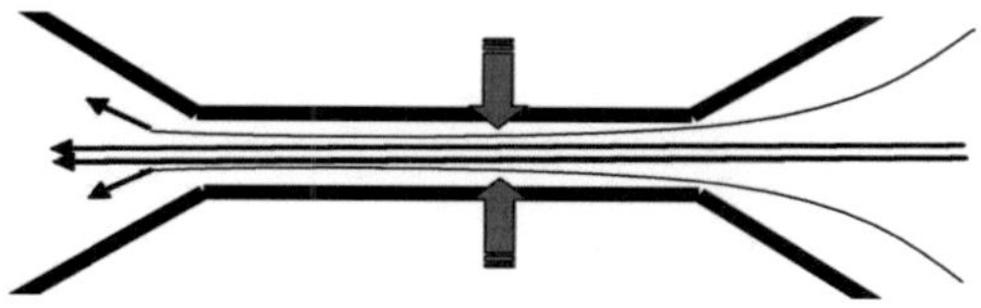

Figure 8.
Venturi effect (author).

2. Using upper openings in atriums and skylights to improve the performance of the effect

3. Using elements like ventilation ducts and pipes to improve the performance of the effect

3.3.4 Wind tower (wind catchers) and solar chimney

Wind towers are used to catch air from higher levels and push it into the interior spaces of a building. A cooling process takes place by heat exchange between the walls of the tower and the hot collected air or by using evaporative cooling at the bottom of the wind tower. Fresh cold air flows to the inner spaces through an opening at the end of the wind tower. At night, wind tower works as a chimney to suck the hot and exhausted air from the room to the outside environment, causing cooler air to replace hot air from other openings. The performance of wind towers can be improved by implementing a water source, like a fountain, at the bottom of the wind tower, which helps cool the gathered air. Additionally, the wind tower can be combined with courtyards and underground tunnels to increase the cooling process of the collected air. A wind tower operates in various ways depending on different factors like the time of day, the presence or absence of wind, and the difference of air temperature inside and outside the building (**Table 5**). The fundamental principle of wind tower operation system lies in changing the temperature of the air inside the tower, therefore changing the density, which is a key factor in circulating air and improving the device's performance.

Solar chimney helps to increase the airflow from interior to upper level and to be replaced by cold air from outdoor shaded area like courtyards or basements.

3.4 Slowing

Slowing, as a passive cooling action, refers to the reduction of heat transfer through the building's surfaces by conduction. It depends on the interaction of the building's envelope with the outdoor environment by receiving and absorbing heat and then transferring it to the inner spaces. The performance of slowing as a cooling action depends on many key factors in the building's envelope, such as thermal insulation, thermal masses, building's volume-to-surface area ratio, building materials, and double glazing. These variables control the amount of heat transfer from outdoor environment to inner spaces and therefore reduce the need for heat removal and the associated cooling loads.

3.4.1 Thermal mass

The two main factors designers should take into consideration when choosing a thermal mass material and surface are *thermal time constant (TTC)* and *diurnal heat capacity (DHC)*. These two factors describe the behavior of an area of material when subjected to heat and the time needed to store and release heat. The relative values of TTC are particularly important when the building is affected by a heat flow, while the

Element's construct	Variables	Notes
Device's elements	Orientation	The openings to be orientated toward the wind
	Height	Affects the wind speed entering the tower and building and heat absorption by menials
	Material	Thermal mass material cooled hot air
	Inlet opening	The size of inlet and outlet openings affects the amount of collected winds, its speed in the tower itself, and its speed as it enters spaces
	Outlet openings	
	Integration with inner spaces	Affects the patterns of air distribution inside the building

Table 5.
Wind tower variables (author).

DHC values are important when the solar gain affecting the building is considerable. Both measures indicate the amount of interior temperature swings that are expected from a material based on outdoor temperature. The *thermal time constant* is used to describe the behavior of thermal masses in building envelopes, and it depends on the heat capacity (Q) and the heat transmission resistance (R). In short it represents the effectiveness of the thermal capacity in a building. TTC is calculated for an area by multiplying heat capacity per unit (QA) by the resistance of heat flow of that area (R):

$$TTC = QA \times R \qquad (1)$$

where QA = thickness × density × specific heat and R = thickness/conductivity.

In calculating the TTC per area, (TTCA) for a composite wall, the QA × R of each layer, including the outside and inside air film layers, is calculated in sequence, and they are calculated for each layer from the external wall to the center of the section in question. A high value of TTC means a high thermal inertia of the building, and it results in a low interior temperature swings. The *diurnal heat capacity* is used to describe the building's capacity to absorb the solar energy and to release the stored heat. DHC measure is particularly important when designing a thermal mass that is subjected to direct solar heat gain. It is considered a function of density and thicknesses of material layers, specific heat, and conductivity. The total DHC of a building is calculated by adding the DHC values of each surface. The DHC is a measure of how much cold the building can store during the night in a ventilated building. Design with thermal mass should consider distributing mass to absorb heat near the sources and thereafter to release the heat to start new cycle the next day.

4. Combining of devices and integration with design process

Optimizing the performance of passive cooling devices and techniques can be achieved by identifying their relation to building design process, where these devices can be implemented and integrated with other architectural and cooling elements to accomplish more than one function. This integration encourages designers to take into consideration the implementation of passive cooling devices and techniques as an integrated stage within the designing process, like analysis, planning, and evaluation stages.

and strategies in the early stage (analysis) of the design process give the passive cooling an essential part in the design performance values. Considering such strategies

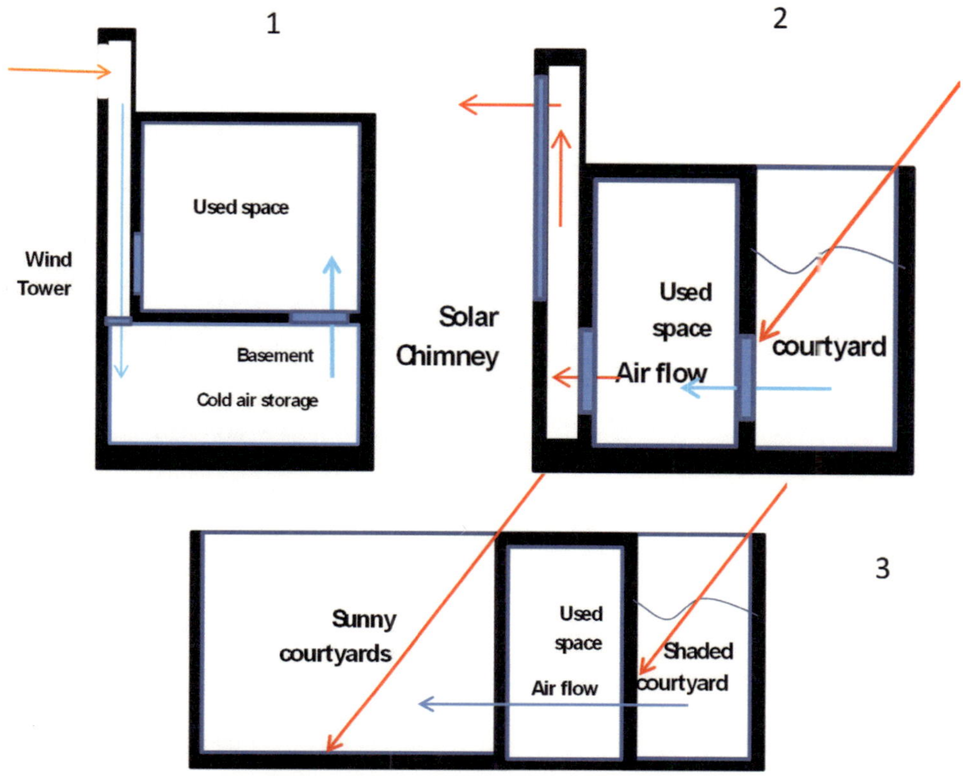

Figure 9.
Example of integration of storage devices and removal devices: (1) combining wind tower with base ment, (2) combining solar chimney with courtyards, and (3) combining two courtyard sunny with shaded one (author).

as design variables will develop building design and create integrated relations between architectural elements and passive cooling devices. Therefore, find creative solutions for passive cooling, and improve the performance of traditional techniques to be easily practiced in modern designs. In addition, considering cooling performance of a building under the designing process as performance criteria in building design process will help designers reevaluate their decisions on passive cooling performance. To do so, computer simulation software can be used to identify which decisions, devices, and variables need to be reviewed during the evaluation stage.

Combining more than one device or principle of passive cooling in building design requires designers to consider passive cooling strategies in all designing processes. The integrated design will create cooling strategies that have a significant and direct impact on building form, plans, sections, and functional distribution, and user's interaction and behavior in the building as seen in (**Figure 9**). Therefore, an integrated building design approach is needed to make the architectural systems, passive cooling systems, and active systems work together within a complete integrated framework to improve performance and save energy, as cooling loads can be minimized through environmental designs that involve judicious use and implementation of shading devices, vegetation, colors, materials, and insulation.

4.1 Design matrix

This paper presents a guideline for implementing passive cooling systems and devices by discussing the four passive cooling actions, which designers should take into consideration in the process of creating a building. It discusses the various

Sources of heat	Actions required	Design solution	Design stage
Direct heat gain from solar radiation on building envelope materials	Avoidance	• Shading devices • Self-shading building form • Landscape design • Urban design	• Early stage (analysis): analyze the relation with local climate, solar angles, and prevailing winds, to prevent heat gain • Middle stage (design): design these variables to protect the inner space from direct gain • Last stage (performance): evaluate the performance of each element in terms of avoidance of heat gain
	Slowing	• Thermal mass • Insulation material • Color	• Early stage (analysis): analyze the relation with local climate, solar angles, and prevailing winds, to slow heat gain • Last stage (performance): evaluate the performance of each element in terms of slowing of heat gain
Direct heat gain from solar radiation on windows and glazed surfaces	Avoidance	• Shading devices • Reflective materials and glass • Orientation design	• Early stage (analysis): analyze the relation with local climate, solar angles, and prevailing winds, to prevent heat gain • Middle stage (design): design these variables to protect the inner space from direct gain • Last stage (performance): evaluate the performance of each element in terms of avoidance of heat gain
	Slowing	• Double glazing	• Early stage (analysis): analyze the relation with local climate, solar angles, and prevailing winds, to slow heat gain • Last stage (performance): evaluate the performance of each element in terms of slowing of heat gain
Indirect heat gain by conduction with outdoor environment through building envelope	Avoidance	• Landscape • Orientation	• Early stage (analysis): analyze the relation with local climate, solar angles, and prevailing winds, to prevent heat gain • Middle stage (design): design these variables to protect the inner space from direct gain • Last stage (performance): evaluate the performance of each element in terms of avoidance of heat gain
	Slowing	• Insulation	• Early stage (analysis): analyze the relation with local climate, solar angles, and prevailing winds, to slow heat gain • Last stage (performance): evaluate the performance of each element in terms of slowing of heat gain
	Removal	• Orientation • Implement removal devices	• Early stage (analysis): analyze the relation with local climate, solar angles, and prevailing winds, to help remove or convey the heat outside inner spaces • Middle stage (design): design these variables to remove hot air from the inner space • Last stage (performance): evaluate the performance of each element in terms of removing gained heat

Sources of heat	Actions required	Design solution	Design stage
Indirect heat gain by convection through the ventilation and infiltration currents	Avoidance	• Controlled ventilation • Ventilation from shaded area	• Early stage (analysis): analyze the relation with local climate, solar angles, and prevailing winds, to prevent heat gain • Middle stage (design): design these variables to protect the inner space from direct gain • Last stage (performance): evaluate the performance of each element in terms of avoidance of heat gain
	Slowing	• Thermal mass	• Early stage (analysis): analyze the relation with local climate, solar angles, and prevailing winds, to slow heat gain • Last stage (performance): evaluate the performance of each element in terms of slowing of heat gain
Internal heat gains by human activities, equipment, machines, and lighting	Removal	• Ventilation	• Early stage (analysis): analyze the relation with local climate, solar angles, and prevailing winds, to help remove or convey the heat outside inner spaces • Middle stage (design): design these variables to remove hot air from the inner space • Last stage (performance): evaluate the performance of each element in terms of removing gained heat
	Avoidance	• Design with daylight	• Early stage (analysis): analyze the relation with local climate, solar angles, and prevailing winds, to prevent heat gain • Middle stage (design): design these variables to protect the inner space from direct gain • Last stage (performance): evaluate the performance of each element in terms of avoidance of heat gain

Table 6.
Design matrix in relation to cooling actions (author).

variables affecting each device within each action, and it explains in details the major issues that need to be considered for each device and action in the three design stages of any building. This integration of passive cooling principles in the design stages represents a new invention in architectural technology and a guideline of passive cooling design for designers and architects. **Table 6** summarizes the required passive cooling actions and design solution that are used to minimize the effects of various heat sources and their implementation considerations in each of the three design stages.

4.2 Integration design: devices and principles for maximum performance

The combination of two principles or devices will be discussed in a way to improve performance, increase efficiency, and integrate devices with building design. Many devices can be combined together to perform more than one function and shift cooling and passive design to be as a comprehensive and integrated design approach.

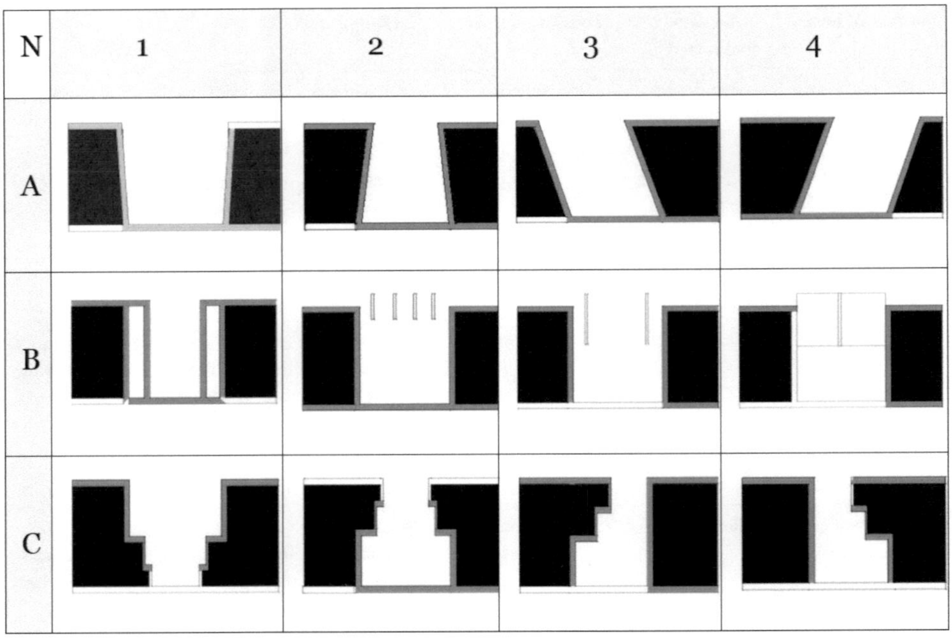

Table 7.
Courtyard configurations [27].

The momentum of passive and green architecture helps to develop new devices that perform more functions and help other devices to perform better.

The design process of such composite design required multi-dimensional analysis of each device and how it could be integrated with other devices.

1. Analysis stage: for each device the working mechanism and condition required to perform well should be thoroughly analyzed.

2. Design stages: design the devices to perform more than one function. In addition design the device to improve the function of other devices

3. Performance stage: reevaluate the integration between devices and architectural systems and how they performed together using experiments or computer simulations.

The following discussion will show how some devices have been integrated and designed with other devices to improve their performance and to become as innovative passive design approaches.

The design of wall geometries of the courtyard could help to improve its cooling performance. Freewan [27] showed that the design of wall geometries helps to control direct incident of sunrays on the courtyard's floor, reduce glare, and improve both daylight quality and quantity. It helped to reduce heat gain from artificial light as it introduces daylight from shaded area. The study showed how wall geometries increase the shading area and time and therefore help to store cold air for long time to ventilate inner spaces with fresh cold air. These configurations as seen in **Table 7** improved courtyard design especially in regions with hot and clear sky [27].

Advanced and modern wind towers were used in university building at the University of Nottingham to be cooling and daylighting devices. They were

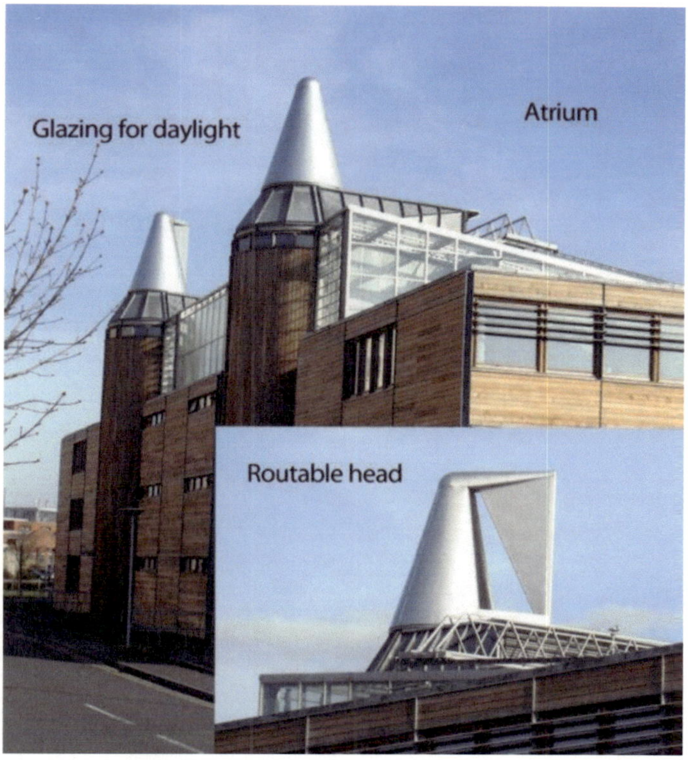

Figure 10.
Combining wind catcher with atrium and daylight elements (author).

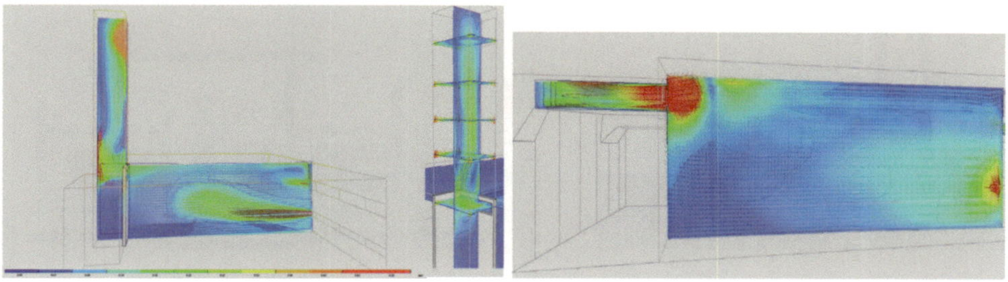

Figure 11.
Integration of windows with wind tower and horizontal duct.

developed to have rotatable head to maximize the efficiency. The new wind towers were designed in integration with atrium and opening figure. The towers were used as wind tower, stairs, daylight devices with large glazed area at top part. They were designed with integration with atrium for maximum performance **Figure 10**.

A study [28] has been conducted at Jordan University of Science and Technology to design adjustable shading devices for existing and new buildings in mild climate with hot summer and cold winter. The research aimed at designing optimized double-positioned external shading device systems that help to reduce energy consumption in buildings and provide thermal and visual comfort during both hot and cold seasons. The design was based on comparison of performance of many variables to determine the best fit characteristics for two positions of adjustable horizontal louvers on south facade or vertical fins on east and west facades for summer and winter conditions. The adjustable shading systems can be applied for new or retrofitted office or housing buildings. The

Form configurations		
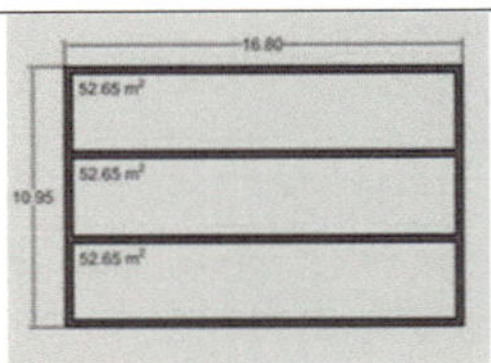 Form 00	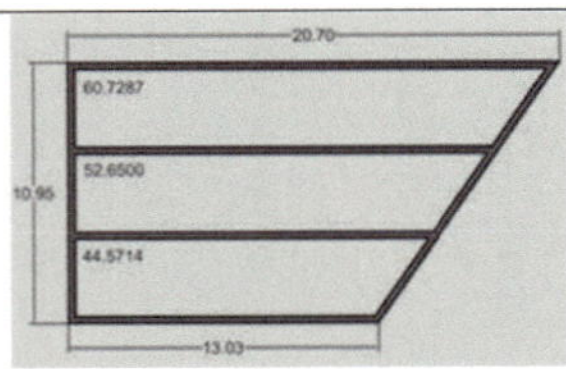 Form 01	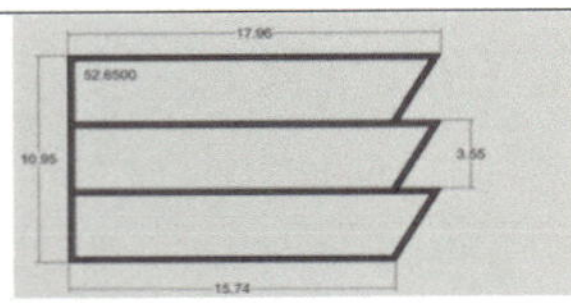 Form 02
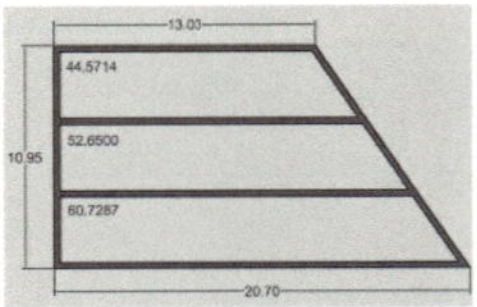 Form 03	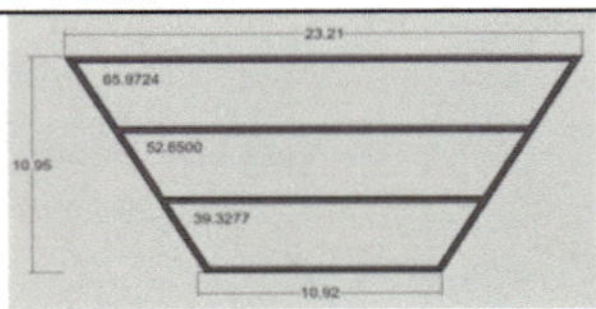 Form 04	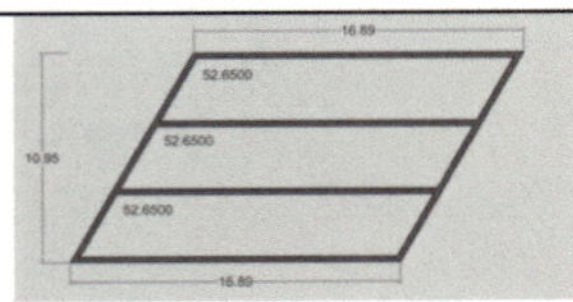 Form 05
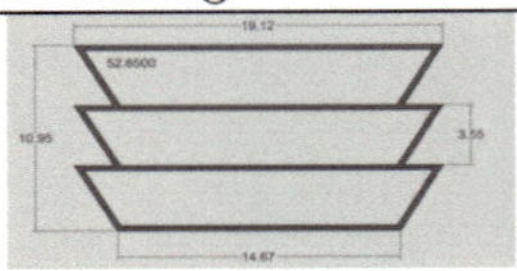 Form 06	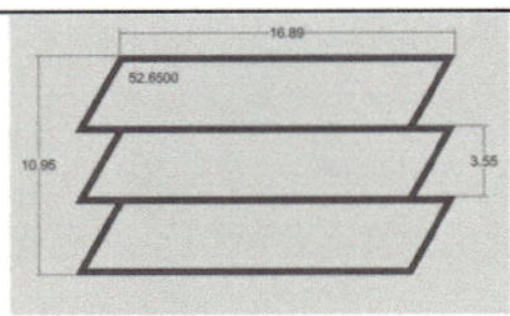 Form 07	

Table 8.
Form configurations [30].

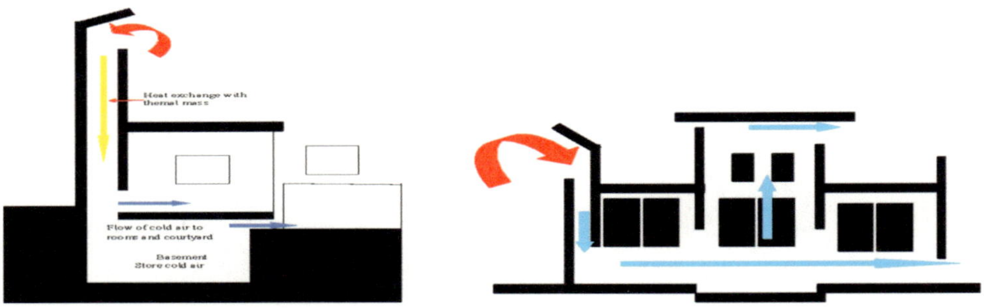

Figure 12.
Combining wind towers with courtyards and basements (author).

optimized shading devices for summer and winter positions helped to reduce the net annual energy consumption compared to a base case space with no shading device or with curtains and compared to fix shading devices.

Freewan and Abdallah [29] studied integration of many devices to improve ventilation process in university classrooms. The study showed that integration of wind tower with side windows or side horizontal ventilation duct with side windows helped to improve the natural ventilation in classrooms, activate stack effect, and increase the air velocity (**Figure 11**).

Freewan [30] studied how the building's form and wall geometries could help to reduce energy consumption and improve thermal and visual comfort. Inward and outward tilted south and north facing facades were studied in the study. Thermal energy performance and daylighting were investigated for many inward and outward angles for both south and north directions. The tilted configurations were achieved as an acceptable balance between cooling, heating energy consumption, and daylighting performance and compared to vertical facades

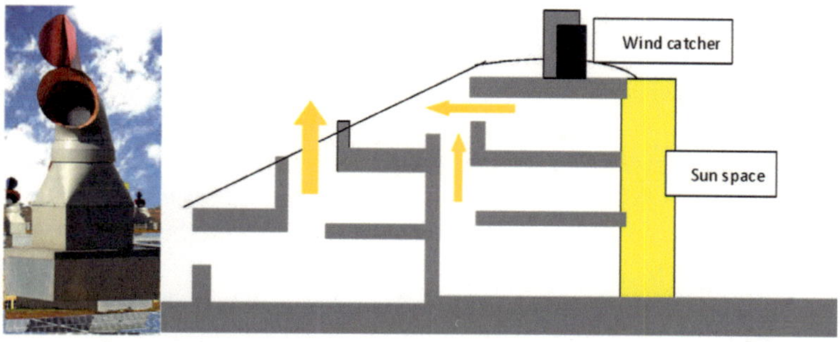

Figure 13.
BedZED design and wind catcher (author).

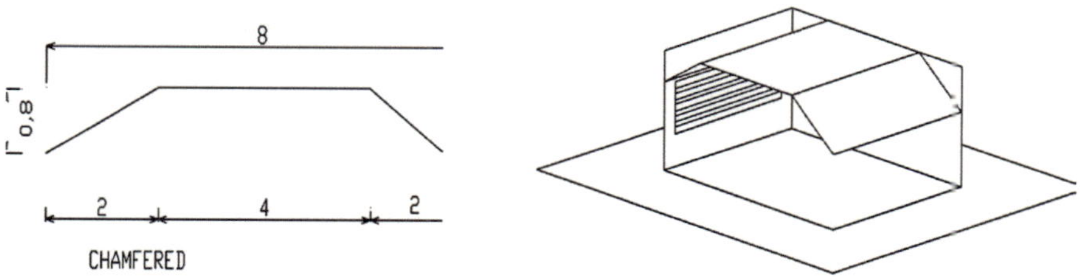

Figure 14.
Combing shading devices with ceiling geometries [31, 32].

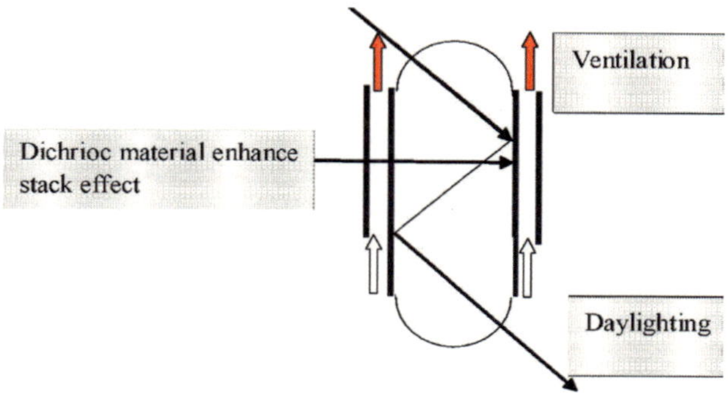

Figure 15.
Working mechanism of light pipe as daylighting devices and ventilation devices (author).

to provide best solar shading, energy consumption, and daylight performance. Many variables were monitored and studied like self-shading, time and period of exposure to sun rays, and how the tilted facade performed. The results showed outward tilted facades for the south orientation performed well as they reduced cooling load and improve both daylight quality and quantity. On the other hand, inward facades for north orientation performed well in terms of daylight compared to vertical facade (**Table 8**).

Wind tower as ventilation and heat removing tools was integrated with courtyard or basement to increase airflow rate and bring cold air to be stored. Configurations like these can be found in Iraq and Egypt, which help activate the stack effect to circulate the cold air to the occupied spaces (**Figure 12**). In modern design wind tower can be integrated with wind tunnel as the wind tower is used to circulate the air, while the tunnel is used to cool the air.

In Beddington Zero Energy Development (BedZED) in the UK, wind catchers with routable head were designed in integration with the buildings' form, sun space, space articulation, and functional zoning (**Figure 13**). The BedZED climatic design was based on more than ventilation principles.

Shading devices and light shelf were studied to be integrated with ceiling geometries in order to maximize shading and daylight performance in hot climate to save energy (**Figure 14**). Many ceiling geometries were investigated to find maximum daylight performance while keeping the optimum shading effects [31, 32].

Light pipe is an advanced daylighting technology used to bring light to a space with no direct contact to outside. It is a cylindrical tube connected to a collecting unit and a diffusing unit. The literature review shows that many researchers have studied the light pipe. Elmualim et al. [33] used dichroic material to develop the light pipe's performance as an integrated system for daylighting and ventilation. The integration is based on using two concentric channels for both daylighting and natural ventilation; the inner one will guide sunlight and daylight into occupied spaces, while the outer one enables passive stack ventilation (**Figure 15**).

5. Conclusions

Implementing passive cooling systems in building design has many advantages over using the fossil fuel-based cooling systems, as they produce no environmental impacts and GHG emissions. The implementation of passive cooling devices in any building design requires many considerations and analyzation of the various variables affecting the cooling performance, and these considerations need to be taken from the earliest design stages and not only at the end of the architectural project, to allow these systems to reach the fullest possible potentials and to be integrated within the design itself, rather than being an additional solution that is forced into a building.

This chapter represented a guideline and innovations in building design process on a comprehensive level that take into consideration the four passive cooling actions, *store*, *avoid*, *remove* and *slow* of heat, and the different devices used for implementing each of the four actions and the variables affecting their cooling performance. All the actions, devices, and variables then were discussed within the three design stages: analysis, designing, and performance stages. The research then concluded with a summary of the required passive cooling actions and the design solutions that need to be used to minimize the effects of the various heat sources and the implemented considerations in each of the three design stages. This chapter encourages designers to integrate passive cooling solutions, actions, and devices in the designing process from early stages of the design while taking into consideration the different variables and requirements concerning the passive devices and their implementation in all design stages.

Author details

Ahmed A.Y. Freewan
Jordan University of Science and Technology, Irbid, Jordan

*Address all correspondence to: aafreewan@just.edu.jo

References

[1] Lechner N. Heating, Cooling, Lighting: Design Methods for Architects. 2nd ed. New York: John Wiley & Sons; 2015. p. 640

[2] Nunes AIF, Oliveira Panão MJN. Passive cooling load ratio method. Energy and Buildings. 2013;**64**:209-217

[3] Kachkouch S et al. Experimental assessment of thermal performance of three passive cooling techniques for roofs in a semi-arid climate. Energy and Buildings. 2018;**164**:153-164

[4] Prieto A et al. Passive cooling & climate responsive façade design: Exploring the limits of passive cooling strategies to improve the performance of commercial buildings in warm climates. Energy and Buildings. 2018;**175**:30-47

[5] Tejero-González A et al. Assessing the applicability of passive cooling and heating techniques through climate factors: An overview. Renewable and Sustainable Energy Reviews. 2016;**65**:727-742

[6] Panchabikesan K, Vellaisamy K, Ramalingam V. Passive cooling potential in buildings under various climatic conditions in India. Renewable and Sustainable Energy Reviews. 2017;**78**:1236-1252

[7] Oropeza-Perez I, Østergaard PA. Active and passive cooling methods for dwellings: A review. Renewable and Sustainable Energy Reviews. 2018;**82**:531-544

[8] Samuel DGL, Nagendra SMS, Maiya MP. Passive alternatives to mechanical air conditioning of building: A review. Building and Environment. 2013;**66**:54-64

[9] Zamani Z, Heidari S, Hanachi P. Reviewing the thermal and microclimatic function of courtyards. Renewable and Sustainable Energy Reviews. 2018;**93**:580-595

[10] Sadafi N et al. Evaluating thermal effects of internal courtyard in a tropical terrace house by computational simulation. Energy and Buildings. 2011;**43**(4):887-893

[11] Meir IA, Pearlmutter D, Etzion Y. On the microclimatic behavior of two semi-enclosed attached courtyards in a hot dry region. Building and Environment. 1995;**30**(4):563-572

[12] Muhaisen AS. Shading simulation of the courtyard form in different climatic regions. Building and Environment. 2006;**41**(12):1731-1741

[13] Berkovic S, Yezioro A, Bitan A. Study of thermal comfort in courtyards in a hot arid climate. Solar Energy. 2012;**86**(5):1173-1186

[14] Muhaisen AS, Gadi MB. Mathematical model for calculating the shaded and sunlit areas in a circular courtyard geometry. Building and Environment. 2005;**40**(12):1619-1625

[15] Aldawoud A, Clark R. Comparative analysis of energy performance between courtyard and atrium in buildings. Energy and Buildings. 2008;**40**(3):209-214

[16] Cho J, Yoo C, Kim Y. Viability of exterior shading devices for high-rise residential buildings: Case study for cooling energy saving and economic feasibility analysis. Energy and Buildings. 2014;**82**:771-785

[17] Palmero-Marrero AI, Oliveira AC. Effect of louver shading devices on building energy requirements. Applied Energy. 2010;**87**:2040-2049

[18] Datta G. Effect of fixed horizontal louver shading devices on thermal

performance of building by TRNSYS simulation. Renewable Energy. 2001;**23**:497-507

[19] Yao J. Determining the energy performance of manually controlled solar shades: A stochastic model based co-simulation analysis. Applied Energy. 2014;**127**:64-80

[20] Freewan AAY. Impact of external shading devices on thermal and daylighting performance of offices in hot climate regions. Solar Energy. 2014;**102**:14-30

[21] Hemsath TL, Alagheband Bandhosseini K. Sensitivity analysis evaluating basic building geometry's effect on energy use. Renewable Energy. 2015;**76**:526-538

[22] Zhou H et al. Harvesting wind energy in low-rise residential buildings: Design and optimization of building forms. Journal of Cleaner Production. 2017;**167**:306-316

[23] Zerefos SC et al. The role of building form in energy consumption: The case of a prismatic building in Athens. Energy and Buildings. 2012;**48**:97-102

[24] Caruso G, Fantozzi F, Leccese F. Optimal theoretical building form to minimize direct solar irradiation. Solar Energy. 2013;**97**:128-137

[25] Azari R et al. Multi-objective optimization of building envelope design for life cycle environmental performance. Energy and Buildings. 2016;**126**:524-534

[26] Yasa E. The evaluation of the effects of different building forms and settlement areas on the thermal comfort of buildings. Procedia Engineering. 2017;**205**:3267-3276

[27] Freewan AA. Modifying Courtyard Wall Geometries to Optimize the Daylight Performance of the Courtyard. 2011. pp. 57-64

[28] Freewan AA, Shqra LW. Analysis of energy and daylight performance of adjustable shading devices in region with hot summer and cold winter. Advances in Energy Research. 2017;**5**(4):289-304

[29] Ahmed AF, Abdallah TK. Improving classroom energy efficiency and thermal comfort with natural ventilation in university classrooms, case study lecture halls complex at JUST. In: El-Shafie DM, editor. MIC2018: Second Memareyat International Conference—Architecture and Urban Resiliency; Jeddah, KSA. 2018

[30] Ahmed AF. Generation of energy and daylight efficient architecture solutions depending on building forms. In: International Conference on Energy, Environment and Economics; Edinburgh, UK. 2018

[31] Freewan AA, Shao L, Riffat S. Optimising performance of lightshelf by modifying ceiling geometry in highly luminous climates. Solar Energy. 2008;**82**(4):343-353

[32] Freewan AA, Shao L, Riffat S. Interactions between louvers and ceiling geometry for maximum daylighting performance. Renewable Energy. 2009;**34**(1):223-232

[33] Elmualim AA et al. Evaluation of dichroic material for enhancing light pipe/natural ventilation and daylighting in an integrated system. Applied Energy. 1999;**62**(4):253-266

Chapter 4

Optimising Energy Systems in Smart Urban Areas

Bohumír Garlík

Abstract

In this chapter, the urban structure will be defined with zero or almost zero energy consumption, followed by pollution parameters. Energy systems are designed as networks of energy-intensive local hubs with multiple sources of hybrid energies, where different energy flows are collected on the same busbar and can be accumulated, delivered, or transformed as needed into the intelligent urban area. For analysis of the purpose function of our energy system, a micro-network of renewable energy sources (RES) is defined by penalization and limitations. By using fuzzy logic, a set of permissible solutions of this purpose function is accepted, and the type of daily electricity consumption diagrams is defined when applying cluster analysis. A self-organising neural network and then a Kohonen network were used. The experiment is to justify the application of new procedures of mathematical and informatics-oriented methods and optimisation procedures, with an outlined methodology for the design of smart areas and buildings with near zero to zero energy power consumption.

Keywords: unit commitment, microgrid, fuzzy logic, cluster analysis, intelligent building

1. Introduction

Efforts to increase energy savings (electricity, heat, water, gas and fuels), reductions of greenhouse gases and the most environmentally friendly approaches lead to the need not only to deal with buildings and elements within the territory as separate entities in the future, but also to try with maximum effectiveness to design the whole area or city region in which these sub-elements, which will interact and communicate with each other. With this approach, it is possible to considerably improve the behaviour of the whole territory, which is also able to react flexibly to situations in the area, for example, current traffic or air conditions.

The way to apply this approach is the concept of smart cities [25], which combines various principles of efficient object design, operations management, especially with significant energy savings and sharing of information into one functional unit. Given the current absence of any method of approach to the creation of smart cities on a global scale, our work aims to outline the basic approaches in individual parts of intelligent city design, that is, urban areas and descriptions of the possible variants of the solution of sub-elements of the area. The aim of the contents of this chapter will be to illustrate the methodology of this issue.

The principles of a smart city can be divided into several areas of human activity:

- Political (city management level)
- Social (city population level)
- Technological (business level)

Due to the comprehensiveness of the concept of smart cities, this chapter will focus on technological areas such as buildings, energy and media, individual and public transport, public space technologies and information, and from these areas, we will be more interested in saving electricity.

The exact and unambiguous definition of the smart area or smart cities has not yet been established on a global scale, but in general it can be said that in order to be considered as smart, all elements and objects contained in that territory must be designed as smart.

So, it does not concern just the building itself but also energy and media supply systems, water supply, waste management, management of all kinds of transport, public lighting and IOT. Instrumentation of the urban system means that the operation of this system can produce data based on key performance indicators, basically making the system a measurable tool and an intelligent metre.

Instrumentation appears to be appropriate to provide urban networks with efficient use of resources, transport and energy services and other public services. Intelligence refers to the ability to use the information gathered to model behavioural patterns and thus to develop predictive models of probable outcomes, allowing for better decision-making and erudite functions. Pilot testing on our experimental intelligent urban area “Rohansky ostrov” (Rohan Island, Prague 8 district) provides information on how to consume electricity more efficiently (we can also focus on water consumption, consumption of heat, natural gas and oil). In our imaginary intelligent area, **Figure 1** shows intelligent instrumentation is widely observed.

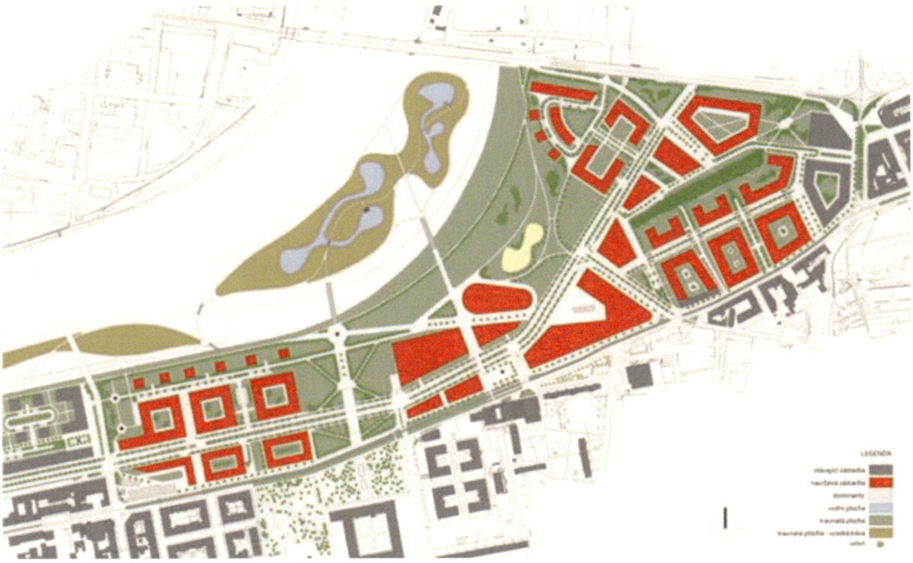

Figure 1.
The architectural design of the locality with newly designed objects in highlighted colour.

Intelligent/smart devices and wireless metres transmit information through broadband networks and provide intelligence that citizens and city organisations can put into practice, thus ensuring its optimisation [29].

For example, in our intelligent Rohan Island area, 250 users can test their energy management system and gain insight into the energy consumption of their appliances, allowing for energy consumption monitoring and remote switching on and switching off appliances. In the intelligent area of the Rohan Island, 500 houses will be equipped with smart energy metres displaying energy consumption Other energy savings have been or are recommended to be discussed in brainstorming sessions. In our Rohan Island project, 500 households will be equipped with smart metres with displays, and personal energy-saving targets will be determined for each household. The goal is to save at least 14% of energy and reduce CO_2 emissions by the same amount. The tallest building in our fictitious smart office area is testing which smart building technology will be best suited to make office buildings more sustainable and more environmentally friendly. Information obtained through smart connections and understanding based on data analysis will be used to provide more effective solutions. In our Rohan Island area—a shopping precinct with many cafes and restaurants and 40 small businesses—solutions for a more sustainable environment will be tested, such as electric vehicle use logistics, energy-saving light bulbs for night light, garbage containers with solar power, smart metres and displays for energy consumption and incentives and benefits from energy savings. A Prague future smart city has recently experimented with crowdsourcing (mass idea exchange of members), that is, it is practically a collaboration since the very beginning of the project, with open innovations, in order to involve its citizens in finding better solutions for public spaces and mobility. Ambitious targets have been set: to reduce CO_2 emissions by 40% and energy consumption by 20% with the implementation of smart zero energy or near zero energy areas by 2025.

In this chapter, we will try to combine the structure of our city—the imaginary intelligent area of Rohan Island—with energy consumption and consequently the pollution parameters. For our experiment, the energy set was chosen as a network composed of energy centers (22/0.4 kV transformer station) with multiple hybrid energy sources where different energy flows are collected on the same busbar and can be accumulated, delivered or transformed as needed. Individual energy centres interact with each other. It is complicated to describe and define it in a comprehensible manner at the municipal level (since it would go beyond the scope of the problem that is dealt with in this chapter). Similarly, it also concerns a challenging generation of new operational models based on existing **critical urban infrastructures**. Critical infrastructure consists of elements or systems of elements (buildings, equipment, resources or public infrastructure) and their operators. Disruption of this function would have a serious impact on the state's security, ensuring the basic living needs of the population, the health of the people or the economy of the state. This is the reason why this issue will be discussed here in terms of assessing the impact on unexpected situations associated with the safety and quality of energy.

The transmission and distribution systems of electricity, natural gas, potable water supply, road and rail transport, communication and information systems and others play an important role in crisis management at the level of cities, urban areas, municipalities and municipalities with extended powers. Therefore, our solution is also focused on specific specifications of the technical and operational values of intelligent information transmission and intelligent networks at the level of the extent of the impact of disruption of their functions. The activities of thermal, electrical and portable infrastructures are considered as qualification characteristics of the energy centre, but they are not taken into account. The experimental part in our case shows that the analysis and optimised layout of the energy system serves one urban district—the urban area (Rohan Island). The associated optimised parametric layout of energy generation infrastructures is a feature of the property of the urban area. An extra vulnerability is due to domino and cascading effects, excessive system complexity and lack of backup. The aim is to protect the information systems (IS) for critical infrastructure (CI), including emergency communication preparedness and protection of materials and equipment which support the IS. For this purpose, a European Programme for Critical Infrastructure Protection (EPCIP) has been set up, and a Critical Infrastructure Warning Information Network (CIWIN) has been built. The European Union is currently planning to increase the protection of Critical Information Infrastructure (CII) in order to ensure the proper functioning of critical infrastructure. The term CII refers to telecommunications, computer systems (including software), the Internet, transmission networks and so on. Nowadays, an especially important component is the Internet, due to its considerable expansion. In our case, the optimisation of energy will be to find solutions for the technical equipment of buildings, such as the internal distribution of engineering and telecommunication networks, starting with the connection to the public distribution of these networks at the level of RES micro-networks. The basic types of energy used in the Czech Republic to produce electricity include thermal, nuclear, solar (sunlight), water and wind.

Many European countries are aiming for a significant reduction in CO_2 emissions by 2050 as well as a reduction in the demand for energy per capita. The European Commission is looking for cost-effective ways to direct Europe's economy towards more climate-friendly and cost-efficient methods. This low-carbon emission economy strategy gives the European Union an incentive to reduce emissions by up to 80% by 2050 compared to levels in 1990. To achieve this, 40% of emissions should be reduced by 2030 and 60% by 2040. All sectors must contribute, and the transition must be appropriate and acceptable; particularly generation and distribution of energy, as well as transport and buildings, are among the main sectors for implementing CO_2 reduction. There are also three main pillars on which the structure of urban energy systems is based [1, 2, 30]. The energy sector has the greatest potential to limit emissions. It can almost completely eliminate CO_2 emissions by 2050. Electricity could actually partially replace fossil fuels in transport and heating. In addition, electricity can be produced with zero emissions using wind, solar, water and biomass energy or other low-emission sources, such as nuclear power plants or fossil-fuelled power plants equipped with carbon capture and storage technologies. This will, however, require high investment in smart grids and micro-network technology [3, 4]. In the short term, the greatest progress can be found for petrol and diesel engines that could be produced with highly improved fuel utilisation and thus more and more efficient. In the longer term, the engagement of hybrid and electric cars will result in a sharp reduction in emissions.

Regarding the European Union strategy planning, emissions from residential and commercial buildings can almost entirely be reduced by approximately 90% by 2050. Energy efficiency will be drastically increased by:

- Passive technologies in new buildings
- Modernisation of old buildings to improve energy efficiency
- Fossil fuel substitutes in the areas of heating, cooling and cooking using electricity and renewable sources of energy (RES)

Electricity begins to play a key role in the smart **urban energy system.** In all existing top examples, the concept of a smart city [6] (or urban area) is based on a recurring cyclical economy and shared resources. Urban energy systems can be seen as a set of energy centres [5], defined as "entities", which uptake energy at entrance ports connected to RES micro-network locations and electricity distribution, and natural gas infrastructures provide certain required energy services, such as electricity, heating, cooling, etc. on the output ports. Inside the centre, energy is transformed and conditioned using, for example, combined energy and heat technology (CHP/FC) transformers, information and communication technology (ICT), compressors, heat exchangers and other equipment. Realistic facilities that can be considered as energy centres include industrial enterprises, larger buildings (hospitals and shopping centres), urban areas and isolated energy systems (trains, trams, etc.). In many cases, other forms of energy to urban areas and vice versa are converted using electricity; thus other forms of energy are generated by electrical energy. From this point of view, the energy system will soon host most energy sources and can be considered a centre of interest for further consideration and in-depth studies.

1.1 Assessment of risks on equipment is expressed in two steps

A relationship (1) for the calculation of risk is defined. This relationship reflects the basic reference variables for the risk calculation, which are the likelihood and severity of the impact of an extraordinary event. In addition, a member taking into account the existing security measures is also included against the classic risk statement. These variables are a function of partial relationships for the calculation of vulnerability, hazards and implemented measures [7]:

$$R = \frac{P \bullet D}{B} = \frac{f(Z_p \bullet N_p) f(Z_D \bullet N_D)}{B} \tag{1}$$

where R is the level of risk, P is the probability of occurrence of an extraordinary event, D is the severity of the impact of an emergency, Zp is the level of vulnerability of the rated equipment that affects the likelihood of an extraordinary event, Np represents the level of threat assessment affecting the probability of occurrence of an emergency, Z_D is the level of vulnerability of the rated equipment that affects the severity of the impact of an emergency, N_D is the level of threat assessment affecting the severity of the impact of an emergency and B is the level of security measures.

In the second step, partial relationships are established to calculate the vulnerability, hazards and workability. The resulting variables of these relationships are a function of the criteria defined in the previous paragraph. These are the following five relationships:

- Rating of the vulnerability level of the equipment affecting the likelihood of an emergency occurrence:

$$Z_p = f(KZ_P, KZ_Z) \tag{2}$$

- Rating of the vulnerability of the equipment affecting the severity of the impact of an emergency:

$$Z_D = f(KZ_K, KZ_O, KZ_R) \tag{3}$$

- Rating the level hazard of threat affecting the likelihood of an emergency occurrence:

$$N_P = f(KN_{PP}) \tag{4}$$

- Rating the hazard level of the threat affecting the severity of the impact of an emergency:

$$N_D = f(KN_A, KN_E, KN_P) \tag{5}$$

- Level of security measures:

$$B = f(KB_U, KB_R, KB_F, KB_C) \tag{6}$$

A weighted arithmetic mean will be used to calculate the individual functions to ensure that all evaluated criteria are adequately represented. At the same time, it should be noted that the criteria related to the assessment of the security level measures include only the newly envisaged security measures. Measures that have already been implemented are reflected in the reduced vulnerability of the facility or reduced likelihood of damage.

1.2 Determining the level of risk

The final step of the critical risk analysis is to define the reference values for determining the resulting level of risk. According to the relationship (1), the level of risk is determined by three variables, namely, the probability of occurrence of an extraordinary event (P), the severity of the impact of an emergency (D) and the level of new security measures (B). Based on this, a 3D model based on the linear shift of the standard risk matrix ($P \times D$) depends on the level of anticipated safety measures (B5 = max., B1 = minimum measures). Using a five-step index scale, all variables reach maximum values of 5 (this ensures the use of arithmetic mean). The resulting risk levels using the five-step index scale are presented in the 3D risk matrix.

1.3 Criteria for assessing the level of vulnerability of the facility

Accessibility (KZ_P)—the ease with which an asset may be affected, whether natural or anthropogenic. Criteria index value: 1–5.

Security (KZ_{FROM})—represents the level of current asset security. Criteria index value: 1–5.

Criticality (KZ_{TO})—the relevance to the system, subsystem or whole component. The objective is critical if its destruction or damage has a significant impact on the performance of the entire system, subsystem, entity or component. Criteria index value: 1–5.

Renewability (KZ_O)—estimates the time needed to replace, repair or bridge the damaged or destroyed asset. Criteria index value: 1–5.

Recognisability (KZ_R)—the time horizon from the origin and identification of the fault after finding its cause. Criteria index value: 1–5.

1.4 Criteria related to threat assessment

Terms of use (KN_{PP})—a set of external factors (such as daytime, climatic conditions and skills) that create favourable or unfavourable conditions for a natural or anthropogenic threat. Criteria index value: 1–5.

Activability (KN_A)—the time horizon of activation of the threat; the longer this horizon is, the less dangerous threat becomes, because there is more time to prepare security measures. Criteria index value: 1–5.

Exposure (KN_E)—the time horizon of exposure to an asset; the longer this horizon is, the more threatening the threat becomes. Criteria index value: 1–5.

Potential (KN_P)—the magnitude of the threat's effect (strength, robustness and yield) is considered by the potential range of impact on the asset. Criteria index value: 1–5.

1.5 Criteria relating to the assessment of the level of security measures

Efficiency (KB_U)—the ability of security measures to minimise the impact of the threat and its impact on the asset. Criteria index value: 1–5.

Feasibility (KB_R)—the availability and usability of technological measures to minimise the threat. Criteria index value: 1–5.

Financial difficulty (KB_F)—the availability of financial resources to implement security measures. Criteria index value: 1–5.

Duration (KB_C)—the time required to implement security measures. Criteria index value: 1–5.

1.6 Municipal energy centres and micro-networks

Within this range, the **urban energy centre** microcosm is one of the most important infrastructures, which is defined as "A group of interconnected loads and distributed energy sources within clearly defined power limits that act as one controllable and manageable entity over the network. The microprocessor can be connected to and disconnected from the network to allow it to operate in both the network connection mode and the Isolated/Autonomous mode". In the CIGRE definition (French: Conseil International des Grands Réseaux Électriques), energy resources are a means of generating and storing resources (heat, etc.). The CIGRE is a leading worldwide community dedicated to the world's knowledge development programme for creating and sharing expertise in energy systems.

In our research, we focused on the goal of understanding the differences between micro-networks and intelligent networks. A microsystem is basically a local island network that can function as a stand-alone or network-connected system. It is powered by gas turbines or renewable energy sources and includes dedicated converters and interconnections to connect to an existing network. Special-purpose filters overcome harmonic problems while increasing the quality and efficiency of electrical power. In short, we are thinking of building a micro-network as a local power provider with limited advanced management tools where the smart grid is a broadband provider with sophisticated capabilities to support automated decision-making. When implementing buildings with zero or almost zero energy consumption, the co-operation of the micro-network of RES with the intelligent network within the 22 kV distribution system takes place. An example of our microsystem that is subjected to our experiment is shown in **Figure 2**.

Micro-networks are the superior physical infrastructure unit that the city's energy centre operates on. If this serves the municipal energy centre [3, 4], then the following issues need to be considered for micro-network activity:

- Independence of urban infrastructures (mobile electricity infrastructure, gas infrastructure, water systems, waste recycling, wastewater treatment)
- Restricted RES penetration, which cannot be considered significant in cities where micro-networks exist (as they are known and defined)

It can be said that the first issues are that infrastructure and urban systems are viewed as individual [7], that is, transport, sewage and water supply, which are usually highly interactive and interdependent (**Figure 3**).

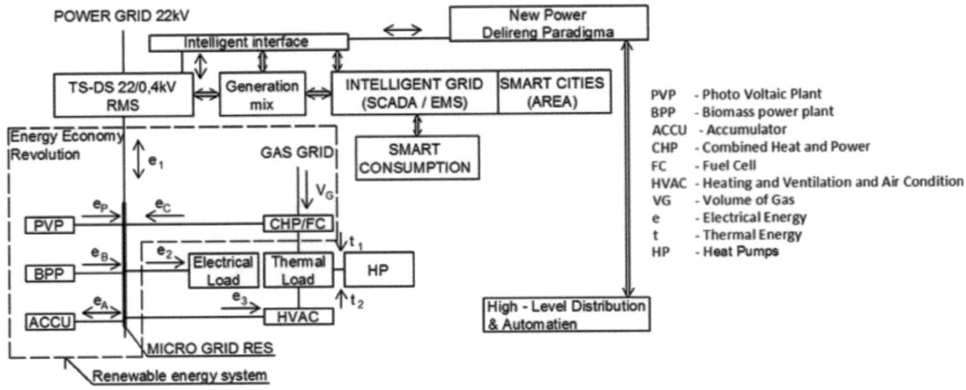

Figure 2.
Microcosm of fictitious RES intelligent regions of Rohan Island [8].

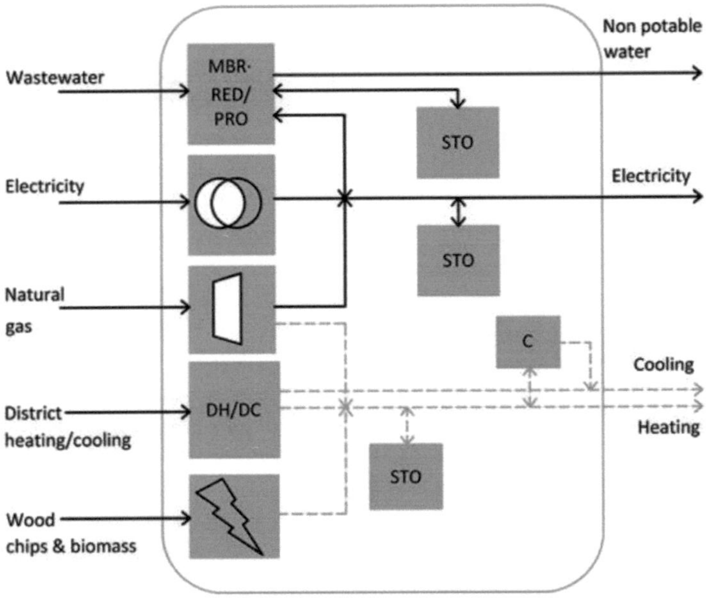

Figure 3.
Example of a municipal power station [9].

2. Unit commitment renewable energy sources and distribution network

In order to ensure the security and reliability of the power supply from RES and the network, electrical resources must be planned and effectively controlled. The large distribution network consists of many elements including generators, transmission lines, transformers and circuit breakers. New RES green energy sources are co-opted or integrated into a distribution system or smart grid system including other DC/DC converters and DC/AC converters and must then be scheduled for microsystem operation. In addition, market structure and real-time energy pricing need to be assessed. For stable operation of the micro-network, it will be necessary to plan electricity generation and supply it to the system load for every second of the intelligent area operation—Rohan Island (Prague 8, Czech Republic) (see **Figure 1**). Energy sources for large energy systems consist of water, nuclear energy, fossil fuels, renewable energy sources such as solar energy (photovoltaic systems) as well as green energy such as fuel cells, biomass and combined heat and power (CHP or also known as cogeneration). These resources must be managed and synchronised to meet the load demand of the microprocessor.

The load requirement of the RES and electricity grids is cyclical and has a peak daily demand for hours and minutes of the week, that is, weekly peak demand for each month and monthly peak demand for the year. **Figures 4** and **5** show the course of electricity demand in our intelligent area on Mondays and Thursdays.

The energy resources must be optimised to meet the peak demand of each load cycle, so that the total cost of generating and distributing electricity is minimised. The power system operator must plan the power sources of the grid and equipment to meet the different load conditions.

Systemic load has a general mathematical formulation. This load gradually increases during the day and then decreases during the night. The cost of the generated power of individual RES sources is not the same for all sources. Therefore, there is a higher effort to produce more energy at the least cost in units. In addition, several network lines connect one electrical network to another neighbouring power grid. These are called interconnections between networks. When exporting power from one power system to an adjacent power supply system through a connecting line, balanced power is considered a load; and conversely, when such power is imported, it is considered energy production. Flow control through these network distributions is

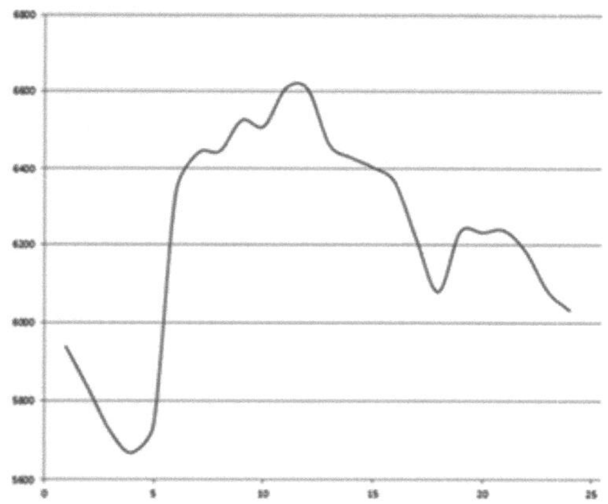

Figure 4.
Electricity demand—Monday.

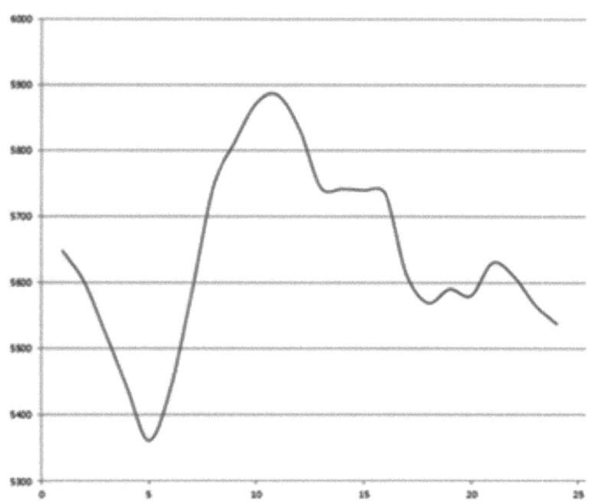

Figure 5.
Electricity demand—Thursday.

preprogrammed (software) based on safe operation and economic indicators. In order to control the energy flow through the connection lines within the transmission at a given frequency of the system, the concept of the control error is introduced (area control error, A_{CE}) and is defined as [10].

$$A_{CE} = \Delta P_n - \beta \Delta f \tag{7}$$

where

$$\Delta P_n = P_2 - P_1$$
$$\Delta f = f_{ref} - f_{mer}$$

and P_2 is the planned power between two power nets; P_1 is the actual power output between two network nets; f_{Ref} is the reference frequency, that is, the nominal frequency; f_{mer} is the actual measured frequency of the system; and β is frequency distortion.

The AGC software (automatic generation control) is designed to achieve the following activities (**Figure 6**):

- Compensation of the surface energy load of the given area, that is, distribution of nodes, links and **load schedule,** thus controlling the system frequency Δf

- Distribution of changing loads between generators **minimising operating costs**

The above conditions are subject to additional limitations that may be introduced in network security considerations, such as loss of power in the line or in the generator.

The first objective is to solve the additional controller and the distortion concept. Parameter β is defined as frequency distortion and is the so-called debug factor that is set when implementing AGC. In the case of a small change in load on the microsystem in the intelligent area, it leads to proportional changes in the system frequency.

For this reason, a bug in a controlled area $A_{CE} = \Delta P_n - \beta \Delta f$ provides each space with information on load changes and controls an additional smart zone controller

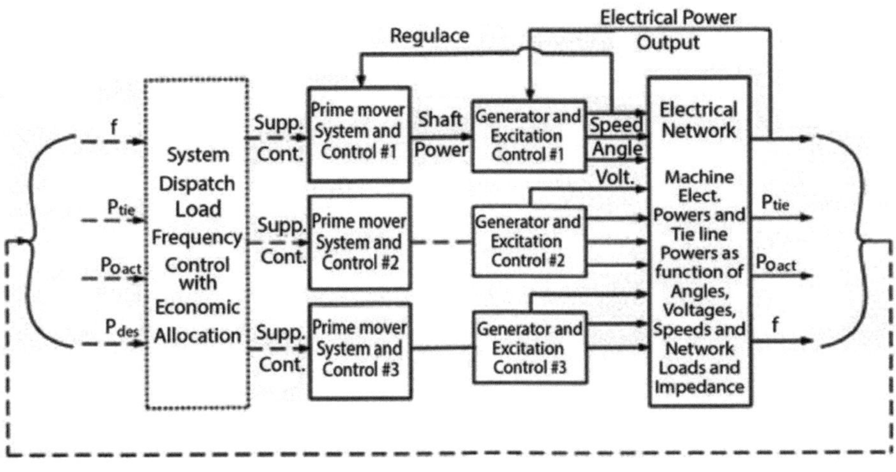

Figure 6.
Control software—Automatic power generator control (AGC) RES [10].

to control the turbine control valves. In order to achieve reasonable regulation (i.e. A_{CE} reduced to zero), system load requirements are sampled every few seconds.

The second objective is to fill energy consumption in the prescribed sample at each minute and to allocate the varying load between the different units to minimise operating costs. This assumes that the load demand remains constant over each period. **Figure 7** shows the AGC block diagram. The AGC also manages the connected micro-networks in a large interconnected power grid. The microgrid concept assumes the grouping of loads in the area within various micro-projects, such as photovoltaics, biomass and combined CHP, acting as a single control network. For the local grid, this cluster becomes the only discernible burden. When the micro-network is connected to the grid, microprocessing voltage is controlled by the local grid. In addition, the frequency of the electrical network is controlled by the operator of the electrical network. The microgrid cannot change the voltage of the bus network and the frequency of the power supply. Therefore, if the microwave network is connected to the local grid, it becomes part of the network and is subject to network failures. The AGC control system is designed to monitor system load fluctuations.

2.1 Economic delivery

The economic supply is expressed in a mathematical process where the required electricity production from the grid including the RES within the micro-network of the intelligent region, Rohan Island, is divided between individual energy sources within the operating RES micro-networks, and thus by minimising defined cost criteria [4], it is subject to both load and operating constraints or penalties.

For each specified load over time (see **Figures 4** and **5**) the power of each RES power plant, including electricity from the distribution grid (i.e. each production unit within the power plant), is calculated to minimise the total cost of fuel required to operate the system load [3]. The problem of economic supply is traditionally formulated as an optimisation with quadratic cost objective functions [11, 24]:

$$f(P_g) = \sum_{i-1}^{N_g} \left(C_i + B_i P_{gi} + A_i P_{gi}^2\right) \tag{8}$$

$$\sum_{i=1}^{N_g} P_{gi} - \sum_{i=1}^{N_D} P_{Di} + P_{loss} = 0 \tag{9}$$

$$s.t\ V_{min} \leq V \leq V_{max}$$

$$P_{gmin} \leq P_g \leq P_{gmax}$$

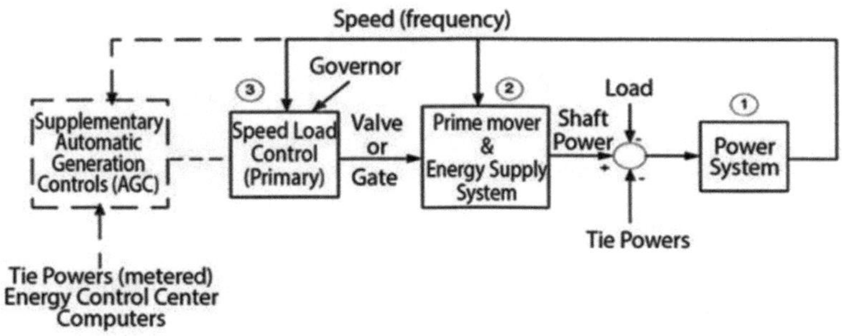

Figure 7.
Control software—Automatic power generator control (AGC) RES [10].

where N_g is the total output produced from all RES, N_D is the total power consumed, P_{zt} is the total power loss in the system, $x_i(t)$ is the energy state of the IT sources over time t (see functions (10), P_{gi} is the power output of the RES, P_{Di} is the power consumed, $P_{gi,\min}$ is the minimum power of the RES source, $P_{gi,\max}$ is the maximum power of the RES source, P_g is the rated power of the RES source, V is the voltage of the RES source, V_{min} is the minimum voltage of the source of RES and V_{max} is the maximum voltage of the RES source.

If independent variable P_g (function argument) P_{gi}, that is, power of the IT resources in time t and $x_i(t)$, is the energy state of the IT source over time t, we get the basic relationship of the cost function:

$$f\left(P_g(t), x(t)\right) = \sum_{i=1}^{N_g} \sum_{t=1}^{T} \left(C_i + B_i P_{gi}(t) + A_i P_{gi}^2(t)\right) \bullet x_i(t) \tag{10}$$

Based on the above-defined variables, constants, mathematical approximations and mathematical structures (8) and (10), we construct two variations of the cost functions of our energy system—the physical model of the intelligent cities' energy system (**Figure 2**). Support is also available [8].

$$f\left(P_g(t), x(t)\right) = \sum_{t=1}^{T} \sum_{i=1}^{N_g} \left(C_i + B_i P_{gi}(t) + A_i P_{gi}^2(t) + \delta_i\left(1 - e^{-\alpha t}\right)\right) \bullet x_i(t) \tag{11}$$

$$f\left(P_g(t), x(t)\right) = \sum_{t=1}^{T} \sum_{i=1}^{N_g} \left(C_i P_{gi} + B_i P_{gi}^2\right) \bullet x_i(t) + A_i P_{gi} x_i(t) \bullet (1 - x_i(t-1) \tag{12}$$

For the sake of our experiment, we will be based on defined cost functions (12) over the entire integrated period (24 hours/day). We will separately allocate the operation and sorting of RES costs. Then we mark the cost function as ***F***, the number of RES in the network is denoted as N_G, the scheduled RES mode (24 hours) will be labelled as $\boldsymbol{t} \in \{1, 2, \ldots, T\}$, the resource index is marked as $\boldsymbol{i} \in \{1, 2, \ldots, N_g\}$, the number of time moments in the given period when RES is depicted as $\boldsymbol{T}$, the power of the IT source at time t, we denote $P_{gi}(t)$, the functional cost of the cost function f is expressed as an algebraic shape $\left(C_i P_{gi} + B_i P_{gi}^2\right) x_i(t)$, the cost of running one of the RES is algebraic $\gamma_i x_i(t) \bullet (1 - x_i(t-1))$, the start-up cost (to commence the operation of RES) in relation (12), is also expressed by the relation $D_i\left(1 - e^{-\alpha t}\right) \bullet x_i(t); \alpha = -\frac{\Delta T_i}{\tau_i}(t)$, the cost coefficients in relation (11) and (12), are C_i, B_i, A_i, D_i, respectively. $\Delta T_i(t)$ and τ_i the relevent cost coefficients, or the downtime and time constant of exponential increas in the start-up costs of the i-th source at time t. $C_i[CZK]$, $B_i[{}^{CZK}/_{MW}]$, $A_i\left[{}^{CZK}/_{MW^2}\right]$, $D_i[K\check{c}]$, the operating costs of RES producing the output will be expressed in an algebraic relationship $P = C + B \bullet P + A \bullet P^2 + D \bullet e$, $x_i(t)$ is the energy state of the IT source over time t and $(x_1, x_2, \ldots, x_n)$, are vector components (independently variable cost functions). Where N_g is the number of resources in the network and T the number of times or slices are considered during the day (24 hours).

We break down the cost (purpose) function (12) and express the operating costs of the RES-generating power $P_g = C \bullet P_{gi} + B \bullet P_{gi}^2 [CZK]$. This relationship is written without expressing the cost. This relationship simplifies the algorithm (12) because the generators (RES) are always in an on state $x_i(t) = 1$. The individual variables (cost items) mean C is cost-dependent on the power output (e.g. the amount of fuel depending on the higher output), *[CZK/MW]*, for example, $C_i \bullet P_i = [CZK]$, and B are costs that are dependent on the second output of power (e.g. joule heating losses are greater with a larger current passing through the conductor). Joule heat is then $Q = RI^2t$ *[J]*. Further, heat is lost in iron and friction $[CZK/MW^2]$ a P_G is the power output *[MW]*.

Restrictive conditions include balances and imbalances according to energy algorithms, as well as generator, bus, voltage and current flow limitations. This is solved through analytical programming, such as nonlinear programming (NLP), quadratic programming (QP) and linear programming (LP), Newton's method, inner point method (IPM) and decision support such as the analytic hierarchy process (AHP). We use alternative methods such as evolutionary programming (EP) [12], genetic algorithms (GA) [13], taboo searching [14], neural networks [15], optimisation of particle flocks [16, 17], the stochastic optimisation algorithm simulated annealing (SA) [18, 29] and adaptive dynamic programming (ADP) which are passive learning methods to improve the performance of the economic delivery algorithm.

2.2 Resource assignment (unit commitment)

Resource deployment, or operational planning function, is sometimes referred to as "pre-delivery". In the overall RES resource management hierarchy [11], resource deployment is coordinated with the planning of economic supply and maintenance and production over time. Scheduling resource deployment covers the scope of the decision on the hourly operation of the power system with a horizon of 1 day–1 week.

Resource planning covers the hourly operation of the RES system with a horizon of 1 day–1 week. We take into account:

a. Restrictions of RES operation and cost per unit of RES resource

b. Restrictions of RES production and reserves

c. Restrictions of running the power plant in terms of RES

d. Restrictions on the local network (micro-networks) RES

While respecting constraints and unexpected stochastic variables, certain assumptions are made when compiling a mathematical statement of resource sorting. These may include, for example, rotary reserves of electricity currents, equipment for respective initial reserves under the conditions of a boiler (in the case of biomass) or partial formulation with the commencement of operation. The **first constraint** is that realistic electricity production must be greater than the sum of the total electricity consumption (power) of consumers in the intelligent Rohan Island area, including required power reserves, or the sum is equal to

$$\sum_{i=1}^{N_g} P_{gi}(t) \geq P_{cel} + P_{rez} \tag{13}$$

P_{cel} is the total required power (net demand) [MW]. a P_{res} is the total power reserve in [MW].

The RES micro-network should maintain a certain power reserve; then the cap of the power reserve must be modified in some way. Hence

$$P_{gi}^{max} = P_{gi}^{cap} - P_{gi}^{res} \tag{14}$$

P_{gi}^{max} is the maximum output power of the IT RES source [MW], P_{gi}^{cap} is the production capacity power of the IT RES source (capacity) in [MW], and P_{gi}^{res} is the production reserve power of the IT RES source in [MW].

$$P_{pop} + P_{zt} \leq \sum_{i=1}^{N} P_{gi} - \sum_{i=1}^{N} P_{gi}^{res} \tag{15}$$

P_{pop} is the production power demand in [MW], P_{zt} is production power loss in [MW], P_{gi} is the total real electricity production in [MW], and P_{gi}^{res} is the total reserve of electricity actually produced

$$A_c = A_0(1 - e^{\alpha t}) + A_L \tag{16}$$

where A_c is the cost of running an off-line resource (resource status in a given hour) in [CZK], a is the thermal time constant, t is the time in [sec], A_L is the workforce costs in [CZK], A_0 is the cost of running the cold boiler in [CZK] and P_{gi}^{max} is the maximum production output power of the IT source [MW].

$$A_{ban} = A_B t + A_L \tag{17}$$

where A_B is the costs to start a subdued resource in [CZK], t is the time in [sec] and A_{ban} are the wage costs in [CZK].

Resource sorting belongs to the classic Lagrange relaxation technique, but the solution to the constraints is based on stochastic variables. That is why we have solved optimisation by simulated annealing as stochastic optimisation (we will not deal with this further; see [8]). Allowed cost functions are conditional when the output power produced by the local network (micro-networks) of RES in the given hour A(t) is determined by the sum of resources turned on. We draw from a typical daily electricity consumption diagram at any given time. The optimisation algorithm works with acceptable solutions (see [8]) which can be evaluated through cost functions without the use of penalties. Then.

$$\sum_{i=1}^{N_g} P_i x_i(t) = A(t), \text{ for } t = (1, 2, \ldots, 24)[\text{h}]. \tag{18}$$

2.3 Optimisation of energy system special-purpose system

The design of special-purpose function $f(x)$ is one of the most complex steps of optimisation. There is no guide or procedure of creating such a function. If we are to design such special-purpose function, we have to know what we are to achieve and what the starting point may be. When we consider our problem, we can see that, in order to achieve reliable and functional results, we have to solve it using constrained optimisation. The constrained optimisation may then be mathematically expressed as follows:

$$\begin{gathered} \textit{minimalise}\, f(x)\ \textit{under restrictive condition}\ g_i(x) \geq 0, i \in, i = k' + 1, \ldots, k \\ h_j(x) = 0, j \in J, j = 1, 2, \ldots, k' \end{gathered} \tag{19}$$

$f : \mathbb{D} \rightarrow \mathbb{R}$, $\mathbb{D} \subseteq \mathbb{R}^d$ is defined above the definition field $\mathbb{D}$, which is a continuous set of searched space, and $\mathbb{R}$ is a real value range. Furthermore, $f, g_i a h_j$ has the functions, and I and J has the final sets of indices. The function F is a special-purpose function; g_i, $i \in I$ has constraints using the inequality algorithm, and

$h_j, j \in J$ has constraints using the algorithm. From a general point of view, the optimisation problem can be expressed as follows:

$$\min f(x)\, for\, x \in \mathbb{R}^n \tag{20}$$

The special-purpose function can be expressed as a sum of quadrates of the deviations between the current parameter values and the required values

$$f(x) = \sum_{i=1}^{m} [y_i(x) - d_i]^2. \tag{21}$$

The value of the minimised special-purpose function or the value of the optimised system parameters depends on the status vector:

$$x = [x_1, x_2, \dots, x_n]^T, \tag{22}$$

where $x_1, x_2, \dots, x_n$ has the state variables of the optimised system expressed by the special-purpose function, $y_1, y_2, \dots, y_m$ has the parameters of the optimised system $d_1, d_2, \dots, d_m$ have the required values of these parameters.

When we introduce the inequality constraint $g_i(x) \geq 0$, the condition expresses that the state variable must be higher than or equal to zero. By multiplying both sides by -1, we get a condition corresponding to the function of state variables. The procedure of its optimisation is then as follows:

$$\min \{f(x) : x \in X\} \tag{23}$$

where $f : X \to \mathbb{R}$ a $X \subset \mathbb{R}^n$.

If our problem is formulated from the point of maximisation, then it is easy to make the adjustment to minimise. In that case, the situation would be the following:

$$\max\{f(x) : x \in X\} = -\min\{-f(x) : x \in X\} \tag{24}$$

$$argmax\{f(x) : x \in X\} = argmin\{-f(x) : x \in X\} \tag{25}$$

For the local minimum, the following applies:

$$on\ X \subset \mathbb{R}^n,\ if\ \ \delta > 0\ so\ that$$
$$for\ each\ y \in X,\ ||y - x|| < \delta\ applies\, f(x) \leq f(y).$$

For the global minimum, the following applies

$$on\ X \subset \mathbb{R}^n,\ if\ foe\ each\ y \in X\ applies\, f(x) \leq f(y) \tag{26}$$

The special-purpose function design is a very complex problem, requiring considerable experience in the subject area, and the possibilities of defining optimisation must be considered. We need to build on what is to be achieved and what can be done.

We have based our experience and our research on optimisation solutions of energy systems from energy companies within the Czech Republic. Based on this we have described the physical model of the energy grid, RES microgrid (**Figure 2**), which corresponds to our experiment. By adjusting the algorithm (18), the relation for **the restrictive condition** of the cost function is gotten using relation (19) and the relation (27). The fact (reality) will be such that $g(x_i(t)) \geq 0$. Then we can write a relationship

$$g(x_i(t)) = \sum_{i=1}^{N_g} P_i x_i(t) - A(t) \geq 0 \tag{27}$$

where $x_i(t) = (x_1(t), x_2(t), ..., x_7(t))$.Dependency $\vec{x}\ (t) = (x_1(t), x_2(t), ..., x_7(t))$ depends on the state of the source at a given hour,

where $\sum_{i=1}^{N_g} P_i x_i(t)$ represents the state of the power generator at time t and $\gamma(t)$ represents the energy consumption forecast for a given hour.
Parameters $i \in \{1, 2, ..., N_g\}$ stand for source indexes. N_G represents the number of sources in our microgrid (which is 7). Variable $t \in \{1, 2, ..., T\}$ represents the time the connected sources spend in the defined mode, and $P_{gi}(t)$ is the output of the source at time t.

2.4 Penalty function

The optimisation algorithm works with acceptable but inadmissible solutions. The penalising function is zero in terms of standard requirements. For one criterion, it has a non-zero value and is positive.

If we add a penalty function to the cost function, then we get an algorithm that is only optimally suitable for local searches in terms of effectiveness. We see this if we exit from [19, 20]; then we can apply a suitable approach to penalising cost functions.

In the first instance Let us define meanings. **Definition 1**: Consider functions f, g, and suppose some values of the function $g\ (x)$ belong to $D\ (f)$. To every such value $u = g(x) \in D(f)$, assign $y = f\ (u) = f\ (g\ (x))$. This defines the function $h\ (x) = f\ (g\ (x))$, which we will call function f, g and mark it $h = f \cdot g$ Note: G is the first function and the second is F. The penalising function is the function of unsolicited power supply:

$$f(P_g(t), x(t)) = (f(X) + a) \bullet \prod_{i=1}^{m} c_i^{b_i} \tag{28}$$

where $x(t) = X = \{x_1, x_2, ..., x_D\}$, $D = 7$ minimises the function $f(P_g(t), x(t)) \equiv f_{cost}(X)$ which is a purposeful function, $c_i = 1.0 + s_i \bullet g_i(X) jestli\ g_i(X) > 0,\ nebo\ c_i = 1\ jinak\ s_i \geq 1, b_i \geq 1;\ \min\ (f(X)) + a > 0$.

The individual parameters have the following meaning: a ensures load function $f(P_g(t), x(t))$ take negative values. Parameter a is set to high. Constant s_i is applied to the functional transformation, and b_i *is* searching for duplicate hypersurfaces. Limited values $g_i(X)$ will be lower than higher for values s_i and b_i. Very often with parameters like $s = 1$ and $b = 1$, the penalty works satisfactorily. This is an external penalty function that links penalties with condition violations. Penalties only apply outside of acceptable solutions. The external penalty is the one that uses exceeding quadrate measures as a penalty [21]. We have a limited minimisation function [22]; then

$$\min f(x); g_i(x) \leq 0, i = 1, ..., m; h_j(x) = 0, j = 1, ..., l,$$

We will replace

$$\min f(X, g) = f(X) + a \sum_{j=1}^{l} h_j^2(X) + a \sum_{i=1}^{m} (g_i)^2(X) \tag{29}$$

where $a = a_1, a_2, ..., a \to \infty$ apr $h_j(x) = 0, j = 1, ..., l;$ we will get

$$\min f(x,g) = f(x) + a\sum_{i=1}^{m} (g_i)^2(x) \quad (30)$$

When applying definition 1, limiting the conditions (27) and the relationship (28) to the target function $fg(X)$ or by applying definition 1 and conditions (27), (29) and (30), we get a modified algorithm [23] as follows:

$$fg(X) = f(X) + ag^2(X) \approx \min \quad (31)$$

where $ag^2(X)$ is the so-called penalty of the non-required electrical power supply. Because g^2 is a negative number, there is a power when $\gamma(t) > \sum_i P_{gi}x_i(t)$. The function value of the target function (31) must be artificially reduced or increased; then w (31) are the same in the algorithm (31).

We are looking for an x allowing us to minimise functions $f(X) + ag^2(X)$.In such case we solve Eq. (31) by minimising cost function *f(X)* and maximising the penalty defined as function $ag^2(X)$. The result of summing the two functions up is the following function, *fg*:

$$fg(X) = f(X) - a\mu(g(X)) \quad (32)$$

We used the minus sign in algorithm (32) as we intend to maximise functional prescription $\mu(g(X))$. Function $g(x)$ defines the output stability of the system, which is why we may set it to zero. This function thus ranges from 0 to 1 and μ is fuzzy zero. Both constraining conditions (31) and (32) may be compared, and the most suitable definition of the constraining conditions may be selected.

When function $fg(X)$ approaches the minimum, which is our intent, we achieve a stable power output balance. Our objective is to minimise both the operation costs and the deviation of production from consumption.

Note: If the power consumption of a given area of "Rohanský ostrov" smart urban area is higher than the production, there will be a minus sign on the right side of Eq. (27) (it will be a negative number). Then we will focus on mathematical expression (28); if the value of expression (27) is very small, that is, close to zero, we have achieved a suitable solution.

Experimental solution:

1. We start from (32), focusing on penalising unsolicited power supply to the smart area from the local RES microgrid. Weight ***a*** (coefficient) defines the conditions of the cost function. In numerical ratio, it is set to such value that the ratio of costs and balances mutually approximately set off.

2. Next, we define the condition of extending the admissible solution.

   a. We set a low permissible deviation between power generation and consumption.

   b. We define the source organisation so that the required output at a given time was as near as possible to the required consumption.

   c. We accept a small permissible deviation which we mark ΔP (**Figure 8**) and accept expression (33), which is the function of the chart showing negative and positive slope.

Let us define the membership function as $X = 0$, which is a classical set, and $\mu A : X \to \langle 0, 1 \rangle$ as the representation [21]. A fuzzy set will then be a

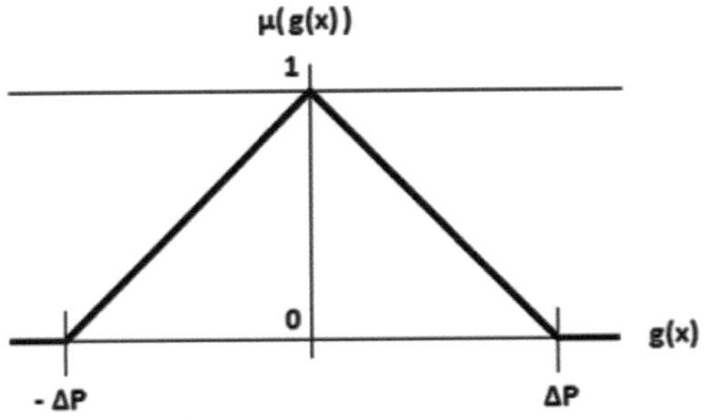

Figure 8.
Maximisation of relation $\mu(g(X))$ by applying the function of triangular fuzzy number zero [8].

coordinated pair $A = (X, \mu A)$. Set X will be the *universe* of fuzzy set A, and μA will be *the membership function of fuzzy set A*. For each $x \in X$, real number $\mu A(x)$ is the level or degree of the membership of element x in fuzzy set A; $\mu A(x)$ will be interpreted the following way:

$\mu A(x) = 0$	Element x is not a member of set A.
$\mu A(x) = 1$	Element x is a member of set A.
$\mu A(x) \in (0, 1)$	It is not possible to identify whether x is a member of A, while the value of $\mu A(x)$ expresses the level, degree of the membership of x in A.

In our case, $g(X)$ expresses the deviation of the stable output balance which is why we seek to set it to zero. Number $x \in X$ is selected arbitrarily from fuzzy set A, and μg is a function of fuzzy set A (where the admissible deviation is defined). It is obvious that for each $x \in X$, real number μ $(g$ $(X))$ may be called the membership level or degree of element x in fuzzy set A.

We describe and compare the expressions $x \equiv g(X)$ and $\mu A(X) \equiv \mu(g(X))]$. Then, $\mu(g(X))$ can be expressed as follows:

$$\mu(g(X)) = (\Delta P - |g(X)|)/\Delta P \text{ in case when } g(X) \in \langle -\Delta P, \Delta P \rangle \quad (33)$$

$$\mu(g(X)) = 0 \text{ in case when } g(X) \notin \langle -\Delta P, \Delta P \rangle \quad (34)$$

Expressions (33) and (34) and **Figure 8** allow us to assume that $\boldsymbol{\Delta P}$=**0**, by which the constraining condition is fulfilled (see $\boldsymbol{\mu((g(X))=1)}$). Eq. (32) is optimised by its subsequent minimisation or maximisation (**μ=0 ǎz 1**). Fuzzy number "μ" will always be small, and we may achieve that using number a (therefore we maximise function $f(X)$). At this moment, we may say that we have solved the optimisation of our task for the purposes of other applications, for example, in order to minimise the special-purpose function.

3. Experiment

Let us assume the fictitious smart city (intelligent area of Rohan Island) which consists of a complex of intelligent residential, administrative and public buildings with a wide range of civic amenities (**Figure 1**).

The energy concept of the area under consideration is clearly focused on local renewable energy sources (FV1, FV2, FV3, FV4 and FV5) assembled together with biomass and cogeneration systems (**Figure 2**), including TS-DS 22/0.4 kV power station, RMS, (**Figure 2**), located in the underground floor of KU02. This is a RES microgrid at a distance of up to 50 km from our fictitious urban area. Continuous and reliable power supply is provided by two high-voltage lines with various switchboards guided from both independent directions. **Table 1** lists the costs, characteristics and technical constraints of individual sources.

Electricity consumption estimates are based on the values of the total usable floor area of all the buildings in the area, and for the estimation of electricity type consumption, specific consumption and consumption values of electricity for months per year for individual types of buildings and the total electricity consumption per year are given in **Table 2**, including financial costs [26–28]. **Table 3** shows

Unit	State	PN	A	B	C
	[Off/on]	mw	[CZK/MW]	[CZK/MW^2]	CZK
FV1	1	140	190	0.50	170
FV2	0	260	190	0.50	230
FV3	0	100	190	0.50	123
FV4	1	50	190	0.50	110
FV5	1	4	190	0.50	95
Biomass	1	1	300	0.40	173
Cogeneration plant	0	1–4	80	0.10	85

Note: Pn is the output rate of the RES-based power plant with a simulation of 0.7.

Table 1.
RES parameters in micro-networks (local RES) [8].

Building type		Adm. And Commercial	Mixed living	Only Living	Cultural	Sport	Total
Floor area Pp[m2]		275 431.00	329 552.00	78 545.00	60 185.00	7 314.00	[illegible]51 027.00
Januar	W_p [kWh/m^2]	2.56	2.64	0.26	4.75	1.08	**11.29**
	P_p x W_p [kWh]	705 103.36	870 017.28	20 421.70	285 878.75	7 899.12	**1889 320.21**
February	W_p [kWh/m^2]	2.23	2.38	0.23	4.21	0.95	**10.00**
	P_p x W_p [kWh]	614 211.13	784 333.76	18 065.35	253 378.85	6 948.30	**1676 987.39**
March	W_p [kWh/m^2]	2.27	2.64	0.25	4.42	0.99	**10.57**
	P_p x W_p [kWh]	625 228.37	870 017.28	19 636.25	266 017.70	7 240.86	**1788 140.46**
April	W_p [kWh/m^2]	2.1	2.56	0.23	4.18	0.93	**10.00**
	P_p x W_p [kWh]	578 405.10	843 653.12	18 065.35	251 573.30	6 802.02	**1698 498.89**
May	W_p [kWh/m^2]	2.06	2.64	0.24	4.19	0.94	**10.07**
	P_p x W_p [kWh]	567 387.86	870 017.28	18 850.80	252 175.15	6 875.16	**7715 306.25**
June	W_p [kWh/m^2]	1.98	2.64	0.23	4.04	0.90	**9.79**
	P_p x W_p [kWh]	545 353.38	870 017.28	18 065.35	243 147.40	6 582.60	**1683 166.01**
July	W_p [kWh/m^2]	2.03	2.56	0.24	4.15	0.93	**9.91**
	P_p x W_p [kWh]	559 124.93	843 653.12	18 850.80	249 767.75	6 802.02	**1578 198.62**
August	W_p [kWh/m^2]	2.06	2.64	0.24	4.19	0.94	**10.07**
	P_p x W_p [kWh]	567 387.86	870 017.28	18 850.80	252 175.16	6 875.16	**1715 306.25**
September	W_p [kWh/m^2]	2.11	2.56	0.23	4.19	0.93	**10.02**
	P_p x W_p [kWh]	581 159.41	843 653.12	18 065.35	252 175.15	6 802.02	**1701 855.05**
October	W_p [kWh/m^2]	2.26	2.64	0.25	4.43	0.99	**10.57**
	P_p x W_p [kWh]	622 474.06	870 017.28	19 636.25	266 619.55	7 240.86	**1785 988.00**
November	W_p [kWh/m^2]	2.34	2.56	0.24	4.45	1.00	**10.59**
	P_p x W_p [kWh]	644 508.54	843 653.12	18 850.80	267 823.25	7 314.00	**1782 149.71**
December	W_p [kWh/m^2]	2.55	2.64	0.26	4.74	1.07	**11.26**
	P_p x W_p [kWh]	702 349.05	870 017.28	20 421.70	285 276.90	7 825.98	**1385 890.91**
Total	**$W_{p,rok}$[kWh]**	**26.55**	**31.10**	**2.90**	**51.94**	**11.65**	**124.14**
Total	**W_{sp}[kWh]**	**7 312 693.05**	**10 249 067.20**	**227 780.50**	**3 126 008.90**	**85 208.10**	**2[illegible]000 757.75**
Total CZK (at distribution rate of D25d)	**CZK 4,10/kWh**	**29 982 041.51**	**42 021 175.52**	**933 900.05**	**12 816 636.49**	**349 353.21**	**8[illegible]103 106.78**

"Note: Wp [kWh / m^2] is the specific electricity consumption per floor area in m^2, Wp, year [kWh / m^2], Wsp [kWh] is electricity consumption per year, PPV [kWp] is photovoltaic power."

Table 2.
Building types and their specific and total consumption.

Building type	RES	P_{PV} [kW_p]	701 x P_{PV} [kWh]	[CZK]
Administrative and Commercial	PS1	2 655.00	1 861 155.00	7 630 735.50
Mixed living	PS2	5 251.00	3 680 951.00	15 091 899.00
Only living	PS3	2 036.00	1 427 236.00	5 851 667.60
Cultural	PS4	1 015.00	711 515.00	2 917 211.50
Sport	PS5	78.00	54 678.00	224 179.80
Distribution area	Biomass	3x1500kWel	22 680.00	47 443.20
Distribution area	Cogeneration	8600 kWel	43 344.00	7 740 000.00
Total	**X**	**11 035.00**	**7 801 559.00**	**39 503 136.60**

Table 3.
Building types and their respective electricity generation per year [8].

electricity generation per year per type of facility including total electricity generation per year.

3.1 Self-organising map

The aim of our experiment is to define a design solution for the system of sorting resources for randomly selected working days Monday and Thursday. Using mathematical analysis using the optimisation stochastic method—simulated annealing—we reached the design and evaluation of input parameters for the purpose of designing the software for the application of the programming language JAVA.

The input parameters for the optimisation programme are hourly load prediction (obtained from the history of experimental scientific observation)—what will be the power consumption at a given time? To evaluate this data, a neural network was used to transmit and process information (data). The neural network was also used to implement and optimise the parameters and structure of the fuzzy model. In addition, the clustering method—cloud analysis of data—was used through data analysis. Several types of daily diagrams were created, and then grouped into "clusters", so that two objects of the same cluster were like two objects from different clusters. The result of the individual clusters was the so-called prototype. Prototypes, cost factors and constraints were input into the neural network, the number of power generators (sources), the number of hours we are functioning on, the cost factors for each generator, the predicted consumption for each hour of the time period and the weight w. A cluster analysis method was applied, and the annual history of electricity consumption has been artificially modelled to compare identified daytime patterns with a standard. The baseline standard used hourly patterns of consumption of the working days on Monday and Thursday in January 2019, where each hourly consumption was randomly modified using a random number generator with a normal probability distribution. This modelling was performed 260 times (the total number of Monday working days) and 260 times (the total number of Thursday working days) through a JAVA programme. In **Figure 9**, two examples of randomly modelled Monday daytime charts are selected, and two examples of randomly modelled daily charts are selected on Thursday. These are hourly consumption forecasts, that is, its standards derived from historical data **Figure 10**.

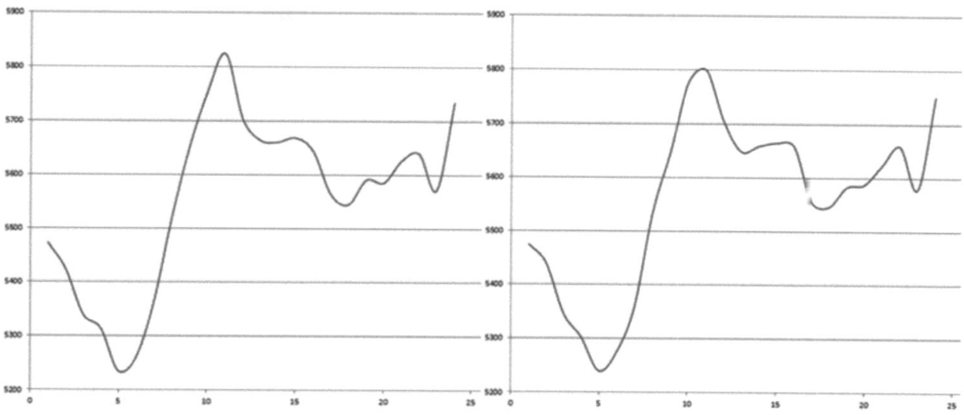

Figure 9.
Typical daily electricity consumption diagram and its standard (Monday).

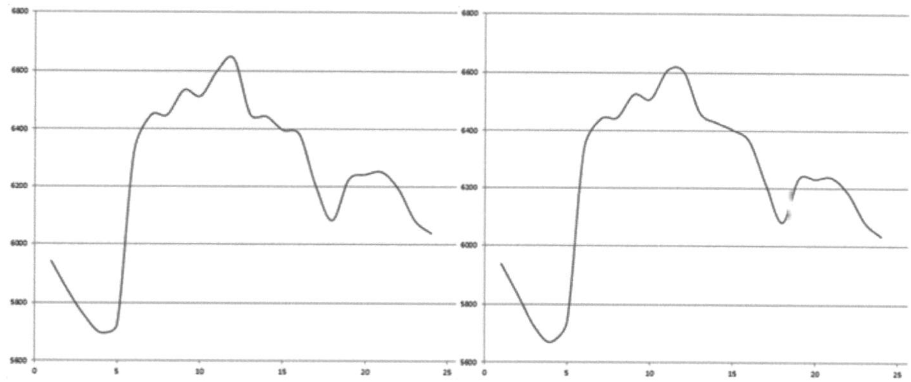

Figure 10.
Typical daily electricity consumption diagram and its standard (Thursday).

We also apply the self-organising neural network (SOM)—the Kohonen network. The Kohonen network works analogously as a cluster or factor analysis. The aim is to reduce the input file by mapping it to a smaller number of clusters. Thus, we can imagine finding the spatial representation of complex data structures so that classes of similar vectors are defined by close neurons in each topology. After the network adaptation, the Kohonen map (**Figure 11**) is drawn out during active

0	0	0	0	0	0	0	0	0	0	0	0	0	0	0
0	0	0	0	0	0	0	0	1	1	0	0	0	0	0
0	0	0	0	0	0	0	4	6	6	5	6	3	0	0
0	0	0	0	2	2	3	3	4	5	3	5	3	1	0
0	0	0	5	5	5	2	4	1	4	1	3	1	2	0
0	0	0	5	8	3	1	3	4	3	4	3	3	1	0
0	0	0	1	1	4	0	9	2	2	11	1	7	1	0
0	0	0	3	5	3	5	4	9	0	8	3	1	0	0
0	0	0	6	5	6	8	5	2	10	3	1	0	0	0
0	0	0	0	0	0	0	0	0	0	0	0	0	0	0
0	0	0	0	0	0	0	0	0	0	0	0	0	0	0
0	0	0	0	0	3	4	5	1	0	0	0	0	0	0
0	0	0	0	0	0	4	0	3	2	4	0	0	0	0
0	3	2	1	0	0	0	6	8	6	2	1	0	0	0
0	3	3	1	1	0	0	2	1	0	0	0	0	0	0
0	7	6	3	2	0	0	0	0	0	0	0	0	0	0
0	3	5	12	0	0	0	0	0	0	0	0	0	0	0
0	0	0	0	0	0	0	0	0	0	0	0	0	0	0

Figure 11.
The Kohonen map of our experiment (cluster analysis).

dynamics, after resubmitting the training patterns, from which we can find a very well-defined massive cluster corresponding to Mondays and Thursdays.

Furthermore, by spreading propagation or active dynamics, we can extract the weight vectors from the configuration of the learned neural network, that is, searched day-type diagrams (**Figures 12** and **13**), where they compare with the appropriate standard.

The individual daily charts in the annual history in **Figures 12** and **13** show that consumption patterns are quite different. Typical daily consumption patterns are basically like the relevant standards (**Figures 12** and **13**), as illustrated by the fact that the cluster analysis method is very effective.

Table 4 shows that the average and maximum tolerances range from approximately 0.1% to 0.5%. From this expression we can evaluate that the cluster analysis method is a very effective and high-quality method demonstrated by this experiment.

If we are to evaluate our experiment according to our specifications, we will assume a situation when we supply power to our smart area of "Rohanský ostrov" through the RES microgrid. The RES microgrid is equipped by eight power generators complemented with the low-voltage grid supplies and the installation of ACCUs. This is a combination of the following ways of power generation:

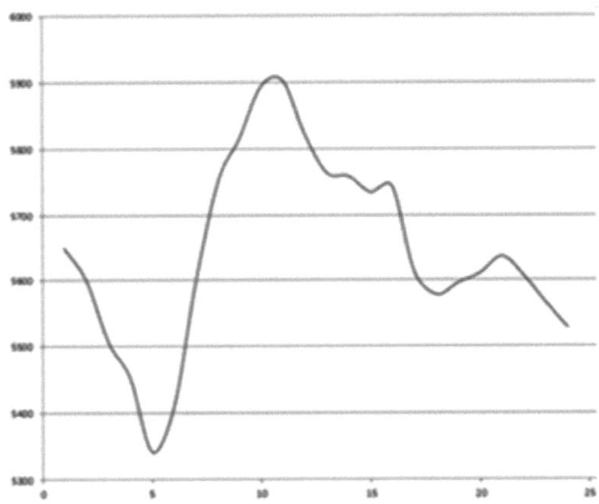

Figure 12.
Typical daily diagram working day Thursday.

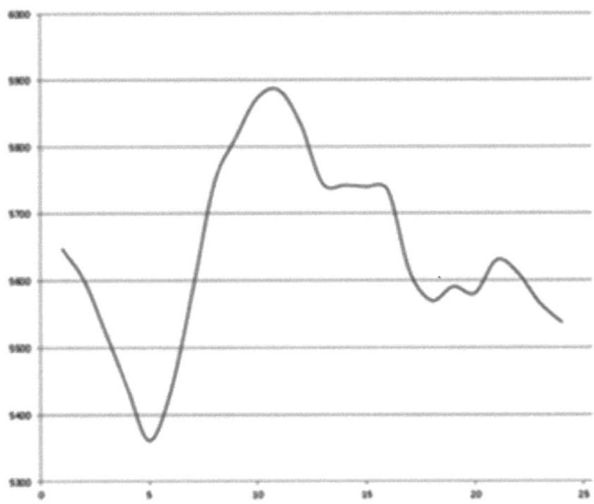

Figure 13.
Standard daily diagram working day Thursday.

Thursday				Thursday			
Time	TDD	SD	Diference	Time	TDD	SD	Diference
(order of hours)	[kW]	[kW]	[%]	(order of hours)	[kW]	[kW]	[%]
1	791	792	0.13	13	2100	2103	0.14
2	650	651	0.15	14	2100	2109	0.43
3	350	351	0.29	15	2000	2007	0.35
4	340	342	0.20	16	1600	1603	0.19
5	1050	1054	0.38	17	1310	1313	0.23
6	2300	2307	0.30	18	1560	1565	0.32
7	2300	2302	0.09	19	1700	1708	0.47
8	2400	2402	0.08	20	1750	1754	0.23
9	2360	2363	0.13	21	1570	1574	0.25
10	2620	2625	0.19	22	1310	1312	0.15
11	2300	2303	0.13	23	1250	1251	0.08
12	2300	2304	0.17	24	1050	1051	0.10
					MEAN=		0.08
					MAX=		0.50

Table 4.
Typical daily diagram (TDD) compared to standard.

photovoltaics, cogeneration and biomass plus low-voltage supplies from the distribution grid (see **Table 5**).

Another task and therefore the aim of the experiment was to design a unit commitment for the weekdays, Monday and Thursday, in January 2019. The hourly consumption forecast has been processed for Thursday (for this chapter we do not specify the hourly consumption of Monday's working day in terms of content) (**Table 5**).

Initial operation of the temperature setting is based on its initial estimate and its subsequent increase to a value at which almost every failure is accepted during the first 10%. The principle of tuning the number of iterations is based on its initial estimate and subsequent increase to a value that does not reduce the resulting production cost to the amount of energy that covers the consumption of that period. The reference cost of electricity generation that covered the estimated consumption of the period was defined as a simplified solution. A simplified solution for that period consisted in the fact that all resources work at medium strength (see relationship (35)).

Unit Commitment of Variant 2

Supply	1	2	3	4	5	6	7	8	9	10	11	12	13	14	15	16	17	18	19	20	21	22	23	24
FV1	0	0	0	0	0	0	0	0	250	300	300	300	300	300	300	300	200	100	0	0	0	0	0	0
FV2	0	0	0	0	0	0	0	200	250	300	300	300	300	300	300	300	200	100	0	0	0	0	0	0
FV3	0	0	0	0	0	0	0	200	250	300	300	300	300	300	300	200	200	100	0	0	0	0	0	0
FV4	0	0	0	0	0	0	0	250	250	300	300	300	300	300	300	250	180	100	100	0	0	0	0	0
FV5	0	0	0	0	0	0	0	200	250	250	300	300	300	300	340	230	200	100	100	100	0	0	0	0
Biomas 1	220	150	100	100	300	820	820	500	500	500	250	250	200	200	150	110	120	360	500	500	500	350	500	410
Biomas 2	250	150	100	100	300	820	820	500	400	450	280	280	210	210	150	100	100	350	500	500	350	340	355	250
Cogenerace	150	150	100	100	400	450	450	500	200	200	260	260	180	180	150	100	100	330	490	450	300	322	400	400
Distribuce	150	150	0	0	0	200	200	0	0	0	0	0	0		0	0	0	0	0	210	410	300		0
ACCU	10	30	30	20	30	0	0	30	0	0	0	0	0		0	0	0	0	0		20	0	0	0
Total [kW]	780	630	330	320	1030	2290	2290	2380	2350	2600	2290	2290	2090	2090	1990	1590	1300	1540	1690	1760	1580	1312	1255	1060
Load [kW]	790	650	350	340	1050	2300	2300	2400	2360	2620	2300	2300	2100	2100	2000	1600	1310	1560	1700	1750	1570	1310	1250	1050
Diff [kW]	-10	-20	-20	-20	-20	-10	-10	-20	-10	-20	-10	-10	-10	-10	-10	-10	-10	-20	-10	10	10	2	5	10

Table 5.
Organisation system for power energy sources of the RES microgrid for the working day Thursday.

$$P_i(t) = C(t)\frac{P_i^C}{\sum_i P_i^C} P_i^C = \frac{1}{2}\left(P_i^{max} - P_i^{min}\right) \tag{35}$$

During the course of optimisation, we perform random settings of the state and output of a randomly selected RES generator within the microgrid using a random number generator. This defined procedure is done for each hour and each set period as well as for every iteration. This is done using a programme in JAVA source code.

```
// start hour cycle
for (int j = 2; j <= nt +1; j ++) {
    // start iteration cycle
    for (iter = 1; iter <= n; iter ++) {
        // state random generation

        // random choice of source
        i = ran (seed) * (ng - 1) +1;
        ij = (i-1) * (nt +1) + j;
        // random change of state
        if (x (ij) == 0) {
                if (ran (seed) <= Ponoff) x (ij) = 1;
        Else
                if (ran (seed) <= Ponoff) x (ij) = 0;
        // random choice of source
        i = ran (seed) * (ng - 1) +1;
        ij = (i-1) * (nt +1) + j;

        // random set of power
        p (ij) = rand (seed) * (Pmax (i) - Pmin (i)) + Pmin (i);
}
}
```

nt is the number of hours and *ng* is the number of sources. p*(ij)* the output of the *i*th generator and **the *j*th** hour. Further, *x(ij)* is the state of the *i*th source during the jth hour. P_{min} and P_{max} are power values with their limit in the *i*th generator. Here, *nt* is the number of hours, *ng* is the number of available sources, *p (ij)* is the power of the *i*th source and the *j*th hour, and *x (ij)* is the state of the *i*th source during the *j*th hour. Values *Pmin* and *Pmax* have their limit in the *i*th source. Ponoff is the parametrizable probability of a change of the source state. Function *ran* is set up by the random number generation within the interval of *(0,1)* with even probability distribution. The result of our experiment in terms of source organisation on the selected Thursday as defined by us is presented in **Table 5**. Worth mentioning is also the fact that the calculation time when using a laptop was 2 minutes and 30 seconds.

4. Conclusion

The total power load (consumption) of the intelligent urban area "Rohanské nábřeží" (Rohan Island) according to **Table 2** is estimated at 21,000,757 kWh/year = 21 MWh/year. Total power generation from RES microgrid (**Table 3**) is 7,801,559 kWh/year = 7.8 MWh/year. At present $\beta = 0.6$ of the total power consumption of the smart area which is 12,600,454 kWh/year. The installed distributed micro-network of RES will cover the power consumption of the urban smart area with 62% of electricity. The projected planned concept (ideal idea) is to have by 2020 a factor of 0.2, thus existing distribution rates can be optimised. In our experiment, 4.2 MWh/year of surplus power would be 85%, which is 3.6 MWh/year. The intelligent urban area would be self-sufficient in terms of electricity consumption and would also generate 3.6 MWh of electricity per year into the 22-kV power grid. The smart area would be energy-efficient in this case, and 85% of the total volume of electricity produced would be commercial. With the transition to smart grids (**Figure 2**), it is assumed that the intelligent urban development of the Rohan embankment will behave like a power producer and be able to influence the energy market. Similarly to the today's use of automated exchange system to offset exchange rate differences, a decentralised network of autonomous buildings —power stations—can be created on the energy market.

When defining the unit commitment optimisation from RES by working day (Thursday) in 1-hour increments, we have achieved a further saving of approximately 20%.

Author details

Bohumír Garlík
Department of Microenvironmental and Building Services Engineering, Czech Technical University in Prague, Prague, Czech Republic

*Address all correspondence to: bohumir.garlik@fsv.cvut.cz

References

[1] Grubler A, et al. Urban Energy Systems. 2014. Available at: http://www.iiasa.ac.at/web/home/research/Flagship-Projects/Global-Energy-Assesment/GEA_Chapter18_urban_lowres.pdf

[2] Evins R, Orehounig K, Dober V. Approaches to Integrated Urban Energy Modelling to Support the Swiss Energy Strategy 2050. 2015. Available at: http://infoscience.epfl.ch/record/213427/9_EVINS.pdf

[3] Gueberro JM, Matas J, De Vicuña LG, et al. Hierarchical control of AC and DC current microcircuits controlled by the constant deviation of the controlled variable—General approach towards standardization. IEEE Transactions in Industrial Electronics. 2013;**58**(1): 158-172

[4] Gueberro JM, Loh P, Chandorkar M, et al. Advanced control architectures for intelligent micro-networks—Part I: Decentralised and hierarchical control. IEEE Transactions in Industrial Electronics. 2013;**1**(1):1254-1263

[5] Geidl M, Koeppel G, Favre-Perrod V, et al. "Energy center—Powerful concept for future power systems". In: Carnegie Mellon's Third Annual Conference on Electrotechnical Industry, 13-14 March 2007

[6] Riva Sanseverino E, Riva Sanseverino R, et al. Smart Rules for Smart Cities: Effective Management of European Cities in the Mediterranean Countries. Springer; 2015

[7] Řehák D, Čígler J, Němec P, Hadáček L. Critical Infrastructure of Electroenergy Determination, Assessment and Protection. 2013. Available from: file:///C:/Users/bohumir/Downloads/Kritikinfrastrukturaelektroenergetiky_urovnposuzovnaochrana.pdf. ISBN: 978-80-7385-126-2

[8] Garlík B. Application of neural networks and evolutionary algorithm to solve energy optimisation and unit commitment for a Smart City. Neural Network World. 2018;**20**(4):379-413. Two: 10.14311/NNW.2018.28.022. ISSN: 1210-0552

[9] Edited by MASSIMO La Scala from Smart Grids to Smart Cities. Wiley. ISBN: 978-1-84821-749-2. 2017

[10] Keyhani A, Marwali M. Smart Power Grids 2011; 2012. ISBN 978-3-642-21578-0, ISSN: 1612-1287

[11] Momoh J. Smart Grid Fundamentals of Design and Analysis. Wiley, IEEE Press; 2018. ISBN: 978-81-265-58124

[12] Yang H, Yang P, Huang C. Evolutionary programming based economic dispatch with non-smoot fuel cost functions. IEEE Transactions on Power Systems. 1996;**11**:112-118

[13] Walters DC, Sheble GB. Genetic algorithm solution of economic dispatch with valve-point loading? IEEE Transactions on Power Systems. 1993;**8**: 1325-1332

[14] Lin WM, Cheng FS, Tsay MT. An improved Tabu search for economic dispatch with multiple minima. IEEE Transactions on Power Systems. 2002; **17**:108-112

[15] Lee KY, Sode-Yome A, Park JH. Adaptive Hopfield neural network for economic load dispatch. IEEE Transactions on Power Systems. 1998; **13**:519-526

[16] Eberhart RC, Shi Y. Comparing inertia weights and constriction factors in particle swarm optimisation. In: Proceedings of the 2000 Congress on Evolutionary Computation. Vol. 1. pp. 84-88

[17] Eberhart RC, Shi Y. Particle swarm optimisation: Developments, applications and resources. In: Proceedings of the 2001 Congress on Evolutionary Computation. Vol. 1. pp. 81-86

[18] Dekkers A, Aarts E. Global optimisation and simulated annealing. Mathematical Programming. 1991;**50**: 367-393

[19] Lampinen J, Zelinka I. New Ideas in Optimisation - Mechanical Engineering Design Optimisation by Differential Evolution. Vol. 1. London: McGraw-Hill; 1999. p. 20. ISBN: 007-709506-5

[20] Zelinka I. Artificial Intelligence in the Problems of Global Optimisation. Prague: BEN Publishing House; 2002

[21] Haftka RT, Gurdal Z. Elements of Structural Optimisation. 3rd Rev. and Expanded Ed. Boston: Kluwer Academic Publishers; 1992. ISBN: 07-923-1504-9

[22] Kokrda L. Optimisation of building constructions with probability constraints [Ph.D thesis]. Brno: Brno University of Technology Diploma; 2015. Leading Diplomats Work Dr. Pavel Popela. p. 68

[23] Garlík B. The application of artificial intelligence in the process of optimizing energy consumption in intelligent areas. Neural Network World. 2017;**27**(5):415. Institute of Computer Science AS CR. ISSN: 1210-0552

[24] Garlík B, Křivan M. The use of evolutionary algorithms to optimise intelligent buildings electricity supply. Neural Network World. 2013;**23**(5):465. Institute of Computer Science AS CR. ISSN: 1210-0552

[25] Pribyl O, Svítek M. System-based approach to intelligent cities. In: Proceedings of the First IEEE International Smart Cities Conference (ISC2-15); in Guadalajara, Mexico. 2015

[26] Svítek M, Votruba P, Moos P. Towards information circuitis. Neural Network World. 2010;**20**(2):241-247. ISSN: 1210-0552

[27] Vlčková V, Hrube P. Traffic accident system model and cluster analysis in GIS environment. Acta Informatica Pragensia. 2015;**2015**(1):64

[28] Postránecký M, Svítek M. SynopCity—Virtual HUB for smart cities. In: EATIS 2016 Cartagena—Columbia. 2016

[29] Pribyl O, Svítek M. System-oriented approach to smart cities. IEEE Systems Man and Cybernetics Society. 2015. p. 1

[30] Svítek M. Towards complex system theory. Neural Network World. 2015; **25**(1):5-33

Chapter 5

Thermal Analysis of an Absorption and Adsorption Cooling Chillers Using a Modulating Tempering Valve

Jesús Cerezo Román, Rosenberg Javier Romero Domínguez, Antonio Rodríguez Martínez and Pedro Soto Parra

Abstract

The energy consumption for space cooling is growing faster than for any other end use in buildings, more than tripling between 1990 and 2016. The efficient use of energy is important to reduce the consumption of electricity of conventional air conditioning. This chapter presents a thermal analysis of absorption and adsorption chillers for conditioning the airspace in a building, controlling the hot maximum temperature at the generator input with a modulating tempering valve (MTV) programmed in TRNSYS and Excel software. The energy performance of the system was maximized based on the tilt of the solar collector, storage tank specific volume, and input generator temperature. The results showed that 35 and 27 l/m^2 of specific volume is a good choice for absorption and adsorption chiller without MTV, and 23 and 22 l/m^2 were selected absorption and absorption chillers using the MTV at a fixed tilt angle of 7° of the solar collector and selecting a minimum temperature at the generator input of 111 and 109°C for absorption chiller without and with MTV, respectively, and 75°C for adsorption chiller without and with MTV. The use of MTV represented a significant reduction of the heater energy for both chillers, mainly for absorption chiller.

Keywords: adsorption chiller, absorption chiller, modulating tempering valve, TRNSYS, air conditioner

1. Introduction

The use of air conditioner and electric fans in buildings around the world is nearly 20% of the total electricity. This consumption has a trend to grow up due to the demographic growth, and more people will naturally want to live in thermal comfort [1]. Solar cooling is a promising and clean alternative equipment which has the advantage for cooling demand in buildings [2]; however, the heat and mass transfer process between the refrigerant and sorption solution and the number of heat exchangers of this kind of equipment (condenser, evaporator, generator, and sorption) reduce the performance of the thermodynamic cycle. The use of new configurations to improve the efficiency of energy has been an important topic of study to reduce the consumption of fossil fuels [1].

The absorption and adsorption cooling system technologies are heat-activated based on the liquid and solid sorption process, respectively. The absorption is based on the absorption and desorption of a working fluid named refrigerant in an absorbent. The absorption cooling system (ACS) consists of four main components: generator, condenser, evaporator, and absorber components as shown in **Figure 1**. The thermodynamic cycle is described as follows: Heat energy is added to the generator to vaporize the refrigerant from the strong solution (high absorbent concentration). The vaporized refrigerant goes to the condenser where it is condensed delivering an amount of heat ($Q_{COOLING}$). The refrigerant leaving the condenser flows through an expansion valve to reduce the pressure and goes to the evaporator; the refrigerant absorbs the heat of the room ($Q_{AIR\ CONDITIONING}$) vaporizes producing the cooling effect. Then, the generated vapor goes to the absorber where it is absorbed by the poor solution of the absorbent coming from the generator, delivering heat ($Q_{COOLING}$), which is dissipated to the ambient to keep the absorption process at a desirable temperature. Finally, the mixture refrigerant-absorbent is pumped to the generator to restart the cycle.

The adsorption system has similar components of the absorption system. However, the cooling effect can be carried out in separate two basic phases (**Figure 2**). In the first phase when the heat is removed ($Q_{AIR\ CONDITIONING}$) from the chilled water, the refrigerant vapor leaves the evaporator and enters the adsorber, where the refrigerant vapor is adsorbed in the adsorbent bed. The second phase is the desorption process as the result of heating up ($Q_{HEATING}$) the adsorbent bed to release adsorbate (refrigerant) from it. The vapor moves to the condenser, and after throttling liquid refrigerant

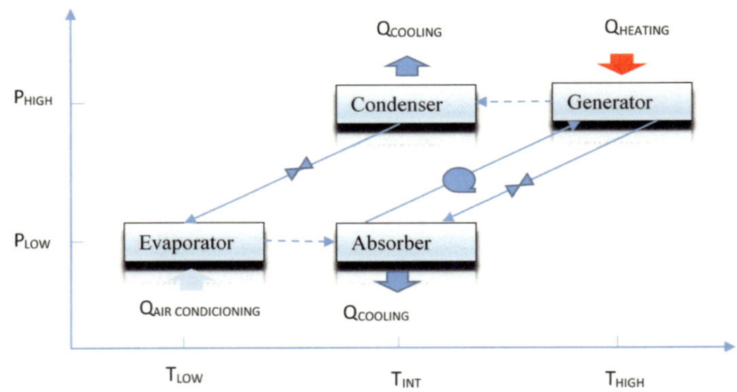

Figure 1.
Schematic diagram of a single-stage absorption cooling system.

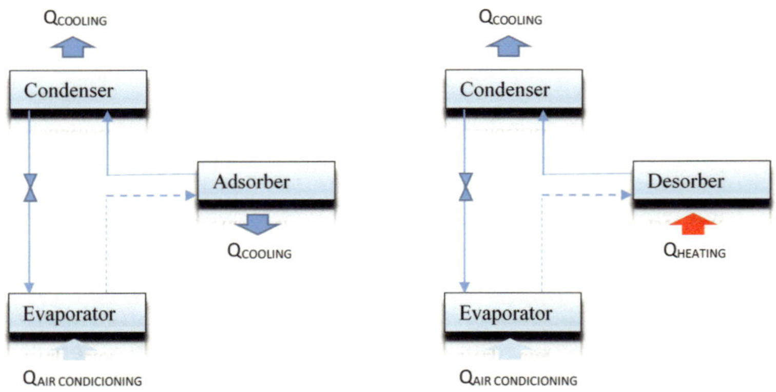

Figure 2.
Adsorption and desorption phases of an adsorption cycle.

	Absorption system	Adsorption system
Generator temperature, °C	85–180	50–100
Coefficient of performance (COP)	0.6–1.1	0.4–0.6
Working fluid	H_2O-LiBr or NH_3-H_2O	H_2O silica gel

Table 1.
Typical operation conditions for the sorption systems.

flows to the evaporator. The adsorption system has two beds to ensure continuous operation which adsorption and desorption processes occur in the same phase as the switching of the chambers [3].

The advantages of the adsorption on absorption chiller are the relatively low temperature to separate the refrigerant in the generator as shown in **Table 1**; however, the coefficient of operation is lower than absorption system [3].

2. Description of the system

The TRNSYS [4] software was used to simulate the solar chillers (SC) for cooling a building located in Temixco City, Mexico. **Figure 3** shows the schematic diagram of the process. There are three main circuits: the solar collector (red line), chilled water (blue line), and cooling (green line). In the solar collector circuit, a heating fluid is pumped (Type 3d-3) from a storage tank (Type 4a) into the evacuated solar collector (Type 71) and returned to the tank. This pump is controlled by Type 2b, and it is turned on when the output is higher than the input temperature of the solar collector or when the temperature of the tank does not reach the maximum temperature of operation in the generator of the SC; otherwise, it is turned off. On the other hand, the heating fluid is pumped (Type 3d-4) to the CC (Type 107–2) to supply energy to the generator. When the temperature is lower than the minimum temperature of the working operation of the SC, a heater (Type 6) is turned on.

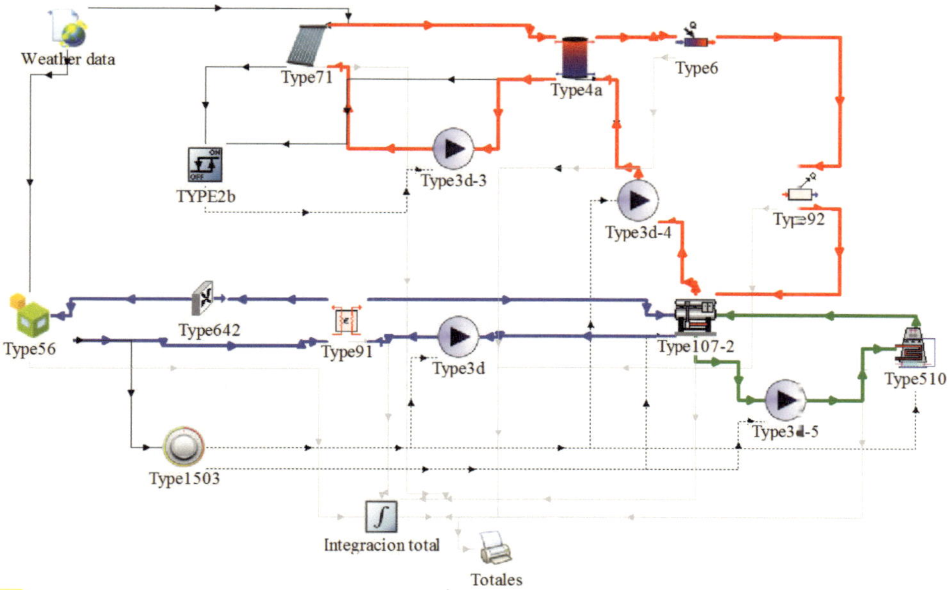

Figure 3.
Single-effect solar absorption cooling system with the auxiliary system in TRNSYS.

The chilled water circuit is used to control the temperature of the building (type 56) using a cooling thermostat (Type 1503) at 25°C ± 1°C. The chilled water is pumped (Type 3d-1) from the evaporator from the SC to a heat exchanger (Type 91) and exchanges heat with an airflow rate coming from the building (Type 642) and returns to the SC.

The cooling water circuit is used to extract the heat load from the condenser and absorber of the SC. A water flow rate is pumped (Type 3d-5) from the SC, and it is sent to a cooling tower (Type 510) which decreases its temperature, and it is returned to the chiller.

Figure 4 shows the implementation of the modulating tempering valve (MTV) programmed in Excel (Type62); in this case, the control Type 2b does not turn off

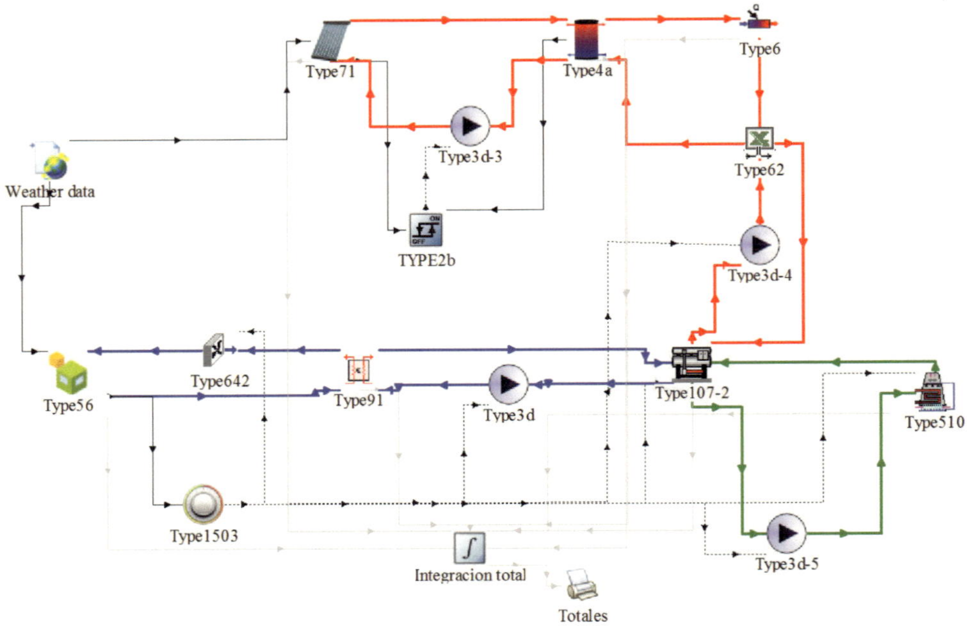

Figure 4.
Single-effect solar adsorption cooling system with MTV in TRNSYS.

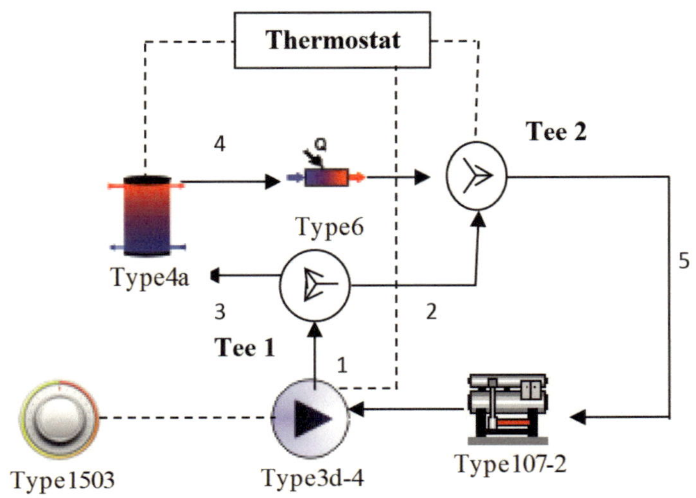

Figure 5.
Connection of the MTV in the solar chiller.

CC	$T_{CHW, set}$	$T_{chilled}$	$T_{cooling}$	T_{hot}
ABS(°C)	6.7	5.5–10.0	26.6–32.2	108.9–116.1
ADS(°C)	6.7	5.0–12.0	10.0–35.0	65.0–95.0

Table 2.
Operation conditions of the CC.

the pump Type 3d-3 when maximum temperature in the generator is reached but when storage tank reaches 130°C.

The MTV consisted of a diverter (Tee1) and a mixer (Tee2) component (**Figure 5**). The objective of the MTV is to keep the input temperature (5) at the maximum range of operation when the storage tank is higher than the limit of operation of CC. The function consists in mixing the high temperature coming from the storage tank (4) with a relative low temperature coming from the chiller (1) changing the flow rates of each stream to keep the maximum temperature (5) at 116°C in the case of absorption system (ABS) and 95°C in the case of adsorption system (ADS) as shown in **Table 2.**

3. Mathematical model of the cooling system

The equations presented in this section describe briefly the main components used for the simulation; a detailed information could be found in [4].

3.1 Assumptions

- The pressure drops and heat losses are negligible into pipelines.
- Fraction pump power (the energy converted to fluid thermal energy, fpar) was 0.05.
- Thermophysical properties of the fluids are constant.

3.2 Basic equations

The solar thermal collector efficiency (Type1b) was calculated using a quadratic or linear efficiency equation according to ASHRAE or European standards. The solar collector is an HTP-Evacuated Tube HP-30SC [5] with a net aperture area of 3.86 m^2. The solar thermal collector efficiency is presented in Eq. (1):

$$\eta = 0.418 - 1.17\frac{(T_{IN} - T_{ENV})}{Q_{RAD}} \tag{1}$$

where T_{IN} and T_{ENV} are the input temperatures of the solar collector and environment temperature, respectively, and Q_{RAD} is the global radiation incident (W/m^2 °C).

The building (Type 56) is a simple multi-zone building with four windows and only one zone with some dimensions and main building characteristics are shown in **Table 3**. Four occupants in a rest position and three computers turned on from 08:00 to 18:00 were considered as heat generation [6].

The Type4a is a stratified storage tank. The properties and working conditions supplied were a heat capacity of 2.24 kJ/kg K for the heating fluid, an initial temperature of 70°C, and a loss coefficient = 0.75W/m^2 K. The characteristics of

Concept	Quantity
North and south wall, m^2	35
Ceiling and floor, m^2	75
West and east wall, m^2	12.5
Thickness of walls (brick), m	0.12
Thickness of insulating (polyurethane), m	0.05
Windows, m^2	2
Air change of ventilation, h^{-1}	6

Table 3.
Building characteristics.

the thermal insulation are density = 16 kg/m^3, thickness = 0.05 m, and thermal conductivity = 0.05 W/m K.

The air conditioning systems used are available in the TRNSYS standard library. This catalog-based component with its own external file predicts the performance of the chiller based on the available range of input data [7]. The file provides values of normalized capacity and COP values as a function of inlet hot water, inlet cooling water, and inlet chilled water temperature.

The absorption (Type107) uses a catalog data lookup approach to predict the performance of a single-effect, hot water-fired absorption chiller. In this design, the heat required to desorb the refrigerant is provided by a stream of hot water. The cooling water stream absorbs the energy dissipated by the absorber and condenser components, and the machine is designed to chill a third fluid stream to a user designated set point temperature. The rated capacity specified was 37,800 kJ/h and a COP rated of 0.70. A thermal fluid was used with a 2.13 kJ/kg °C.

The adsorption chillers (Type 909) cool a liquid stream by evaporating water onto the surface of a solid desiccant matrix. The rated capacity specified was 37,800 kJ/h and a COP rated of 0.53. A thermal fluid was used with a 2.13 kJ/kg °C.

Similar equations are used for both cooling systems. The chilled water, the hot water, and the rejected energy are calculated with Eqs. (2)–(4):

$$m_{CHW}\, Cp_{CHW}\, (T_{CW,IN} - T_{CHW,SET}) = MIN(Q_{CHW}, Q_{CAPACITY}) \quad (2)$$

The energy of hot water.

$$Q_{HW} = m_{HW}\, Cp_{HW}\, (T_{HW,IN} - T_{HW,OUT}) \quad (3)$$

The energy rejected (Q_{CW}) to the cooling water stream.

$$Q_{CW} = Q_{CHW} + Q_{HW} + Q_{AUX} \quad (4)$$

where Q_{HW} and Q_{AUX} are the energy of the hot water and the auxiliary equipment (pumps). $T_{CW,IN}$ and $T_{CHW,SET}$ are the temperature of the input cooling water and the set point of the chiller.

The heater system (Type 6) is used to increase the temperature until the minimum generator temperature of the CC. It was considered to have an efficiency of 100%:

$$Q_{HEATER} = m\, Cp\, (T_{OUT} - T_{IN}) \quad (5)$$

Type 510 models a closed circuit cooling tower; a device used to cool a liquid stream by evaporating water from the outside of coils containing the working fluid.

$$Q_{DESIGN} = m_{WATER}\, Cp_{WATER}\, (T_{IN,DESIGN} - T_{OUT,DESIGN}) \tag{6}$$

$$h_{AIR}(T_{AIR,OUT,DESIGN}) = h_{AIR}(T_{AIR,IN,DESIGN}) + \frac{Q_{WATER,DESIGN}}{m_{AIR}} \tag{7}$$

The heat exchanger (Type 91) relies on an effectiveness (ε) minimum capacitance approach to model a heat exchanger. Under this assumption, the user is asked to provide the effectiveness of the heat exchanger and inlet conditions:

$$QHX = \varepsilon\, Qmax \tag{8}$$

where Q_{MAX} is the maximum heat transfer rate across the exchanger.

The pumps compute a mass flow rate using a variable control function ($0 \le \gamma \le 1$) and a fixed (user-specified) maximum flow capacity (m_{MAX}).

$$m_{IN}\, Cp\, (T_{OUT} - T_{IN}) = P^{*}\, fpar \tag{9}$$

$$m_{OUT} = \gamma\, m_{MAX} \tag{10}$$

where fpar is the fraction pump power, p is the power (kW), m_{OUT} is the output mass flow rate (kg/s), and Cp is the heat capacity of the fluid (kJ/kg°C). **Table 4** shows the input data supplied to the pumps.

The cooling thermostat (Type 1503) is an N-stage cooling aquastat/simple thermostat modeled to output on/off control functions that can be used to control a fluid cooling system having up to an N-stage heating source. The desired fluid temperature may depend on the time of day or the day of the week. These variations of the cooling set point temperatures are modeled using an optional setup control function and a setup temperature difference [4].

The differential controller (Type 2b) is an on/off differential controller that can have a value of 1 or 0. The value of the control signal is chosen as a function of the difference between the upper (T_{HIGH}) and lower (T_{LOW}) temperatures, and both are compared with two dead-band temperature differences ΔT_{HIGH} and ΔT_{LOW}. The new value of the control function depends on the value of the input control function at the previous time step.

The mass and energy balances of the MTV (type 62) are calculated using equations from 11 to 14, respectively:

$$m_1 = m_5 \tag{11}$$

$$m_3 = m_4 \tag{12}$$

$$m_1\, h_1 = m_2\, h_2 + m_3\, h_3 \tag{13}$$

$$m_5\, h_5 = m_2\, h_2 + m_4\, h_4 \tag{14}$$

where h is the specific enthalpy and m is the flow rate and subscripts corresponding to **Figure 5**.

Auxiliary	Power (kJ/h)	Cp (kJ/kg °C)	m (kg/h)
Pump (chilled water)	1339	4.19	1789
Coil (Type 3d-2)	1339	1.22	2487
Pump (Type 3d-3, solar collector)	2664	2.34	5442
Pump (Type 3d-4, hot water)	1339	2.34	1261
Pump (Type 3d-5, cooling water)	1339	4.19	4230
Cooling Tower (fan)	2013	—	—

Table 4.
Parameter supplied to the pumps of the cooling system.

3.3 Parameters

The energetic performance of the cooling systems can be evaluated using two indicators: solar fraction (SF) and heating fraction (HF). Solar and heating fractions are defined as the amount of energy supplied by solar resources (Q_{COL}) and heater system (Q_{HEATER}), respectively, divided by the total energy supplied (Q_{COL} + Q_{HEATER}).

4. Results

Figure 6 shows the behavior of the COP and room and environment temperature for the absorption and adsorption systems during a sunny day (from 2003 to 2016 h). The coefficient of performance is almost keeping constant for ABS (0.71) caused by the short range of operation of T_{HOT} (see **Table 2**) and from 0.46 to 0.48 for ADS. The room temperature was kept from 24 to 26°C; when room temperature is lower than 24°C, the control temperature is turned off (night).

Figure 7 shows the function of the MTV to keep the generator temperature at the maximum range of operation. When the tank temperature (T_4, from **Figure 5**) remains on the range of operation, all flow rates go to the tank (m_1 = m_3 = 1261 kg/h and m_2 = 0, from 1476.1 to1476.8 h); however when it is higher than 95°C (from 1477.0 to 1486.8 h), m_1 splits into m_3 (m_3 = m_4) and m_2 to keep T_5 at 95°C, mixing a high temperature coming from stream 4 with stream 2 (T_2 = T_1).

Next sections will show the condition to operate the cooling system at improved conditions based on the tilt of the solar collector, activation temperature, and storage tank specific volume (SV) [2, 8, 9]. The period of evaluation was from March to May (1460–3560 h) to operate the cooling system due to the highest ambient temperature in a year in Temixco, Mexico [6], as shown in **Figure 8**, besides there are several days with very poor (2190–2300 h) or null beam radiation.

4.1 Selection of the tilt

Figure 9 shows the energy from the solar collector as a function of the tilt of the solar collector. 7° obtained the maximum solar collector energy (1.265×10^7) during the 3 months evaluated using 30 m^2 of area solar collector and a storage tank of 3 m^3.

4.2 Selection of the activation temperature

Figures 10 and **11** show the COP as a function of the input generator temperature without and with MTV for absorption and adsorption chiller, respectively.

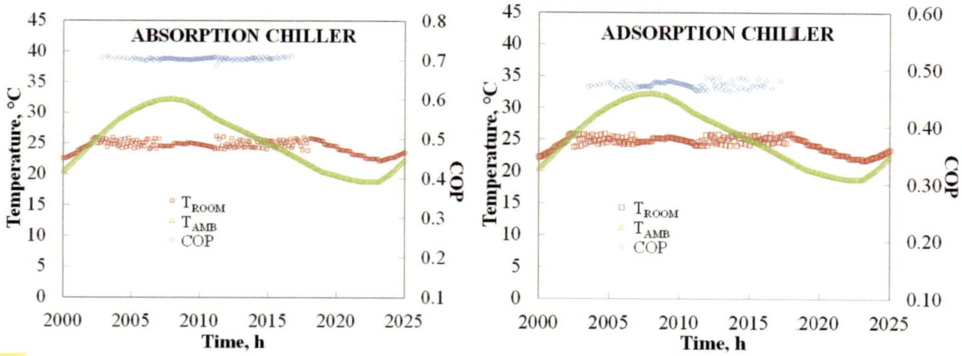

Figure 6.
COP and room and environment temperature behavior within a day.

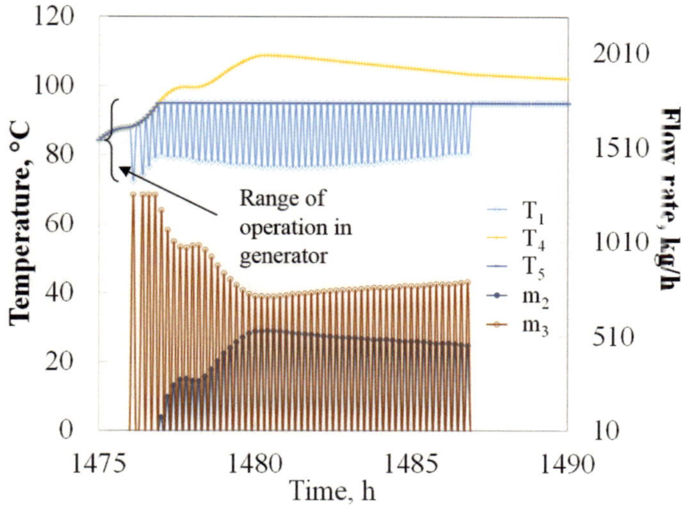

Figure 7.
Behavior of temperatures and flow rates of the modulating tempering valve.

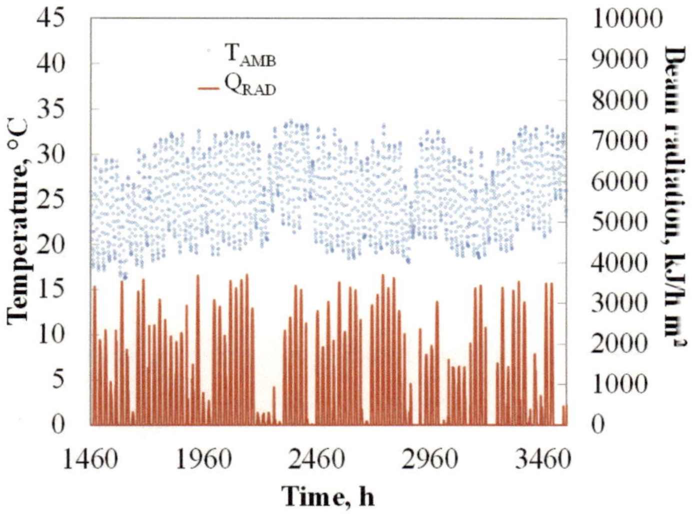

Figure 8.
Environment and beam radiation behavior within 3 months.

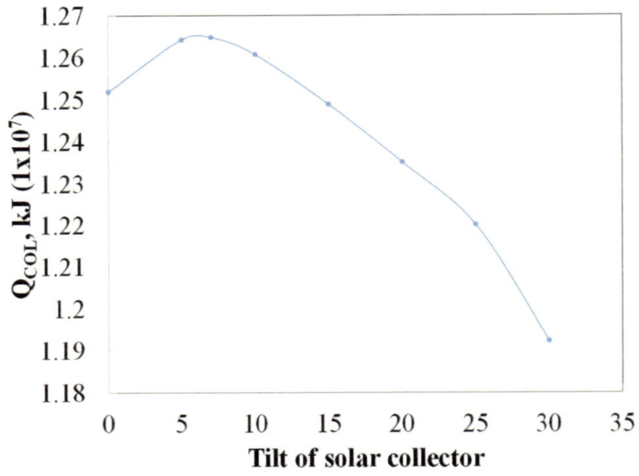

Figure 9.
Behavior of the tilt of the solar collector on solar fraction.

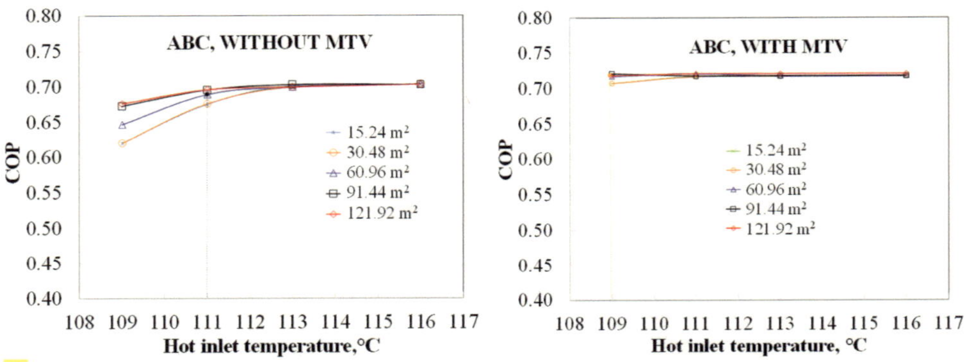

Figure 10.
Effect of the activation temperature on the solar fraction absorption system without and with MTV.

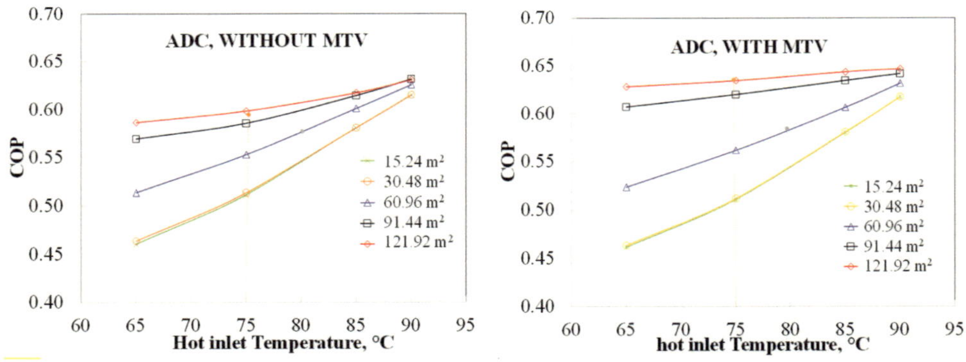

Figure 11.
Effect of the activation temperature on the solar fraction adsorption system without and with MTV.

A temperature of 111 and 109°C was considered a good option to operate the absorption system without and with MTV. While an input temperature generator of 75°C was considered to operate the adsorption system for without and with MTV, 80°C should be used when the area collector is lower than 60.96 m^2.

4.3 Selection of the number of solar collectors and storage tank

The variation of the solar fraction with respect to the specific volume is presented in **Figures 12** and **13** for absorption and adsorption systems, respectively, at different solar collector areas. These figures show that 35 and 23 l/m^2 were a good choice for the configuration without and with MTV for absorption system, respectively, while 27 and 22 l/m^2 were selected for without and with MTV, respectively, for the adsorption system. This represented a 52.2 and 22.7% of reduction for absorption and adsorption systems. A solar collector area of 60.96 m^2 was selected to obtain high values of the solar fraction. Besides a certain storage volume value, the increment of the solar fraction, is not significant [2].

Figures 14 and **15** show the heater energy as a function of the storage tank at several solar collector areas for absorption and adsorption chillers, respectively. It can be seen that there is an increment of heater energy when the volume is incremented at 30.48 m^2 of the solar collector area for both chillers because more energy is required at higher storage volume. However, when the solar collector area is higher than 60.96 m^2, the increment of volume decreases the consumption of energy because the storage tank has a high capacity to keep more energy.

Figures 16 and **17** show the energy of the solar collector (Q_{COL}), heater (Q_{HEATER}), and the total energy supplied (W_{TOTAL}) to the electric equipment

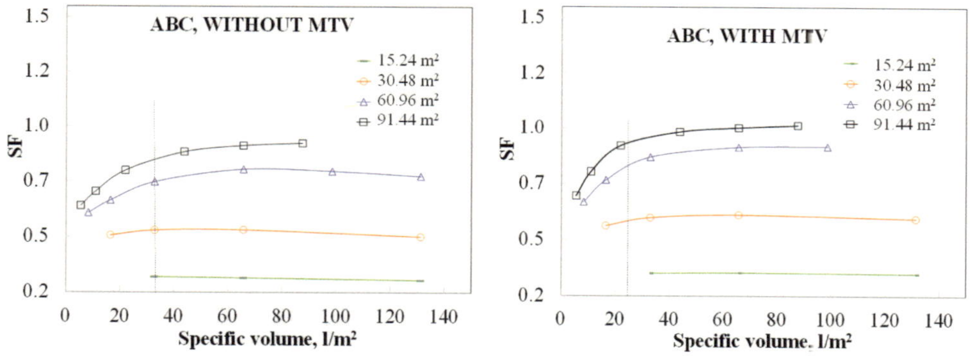

Figure 12.
Solar fraction against storage-specific volume for absorption chiller.

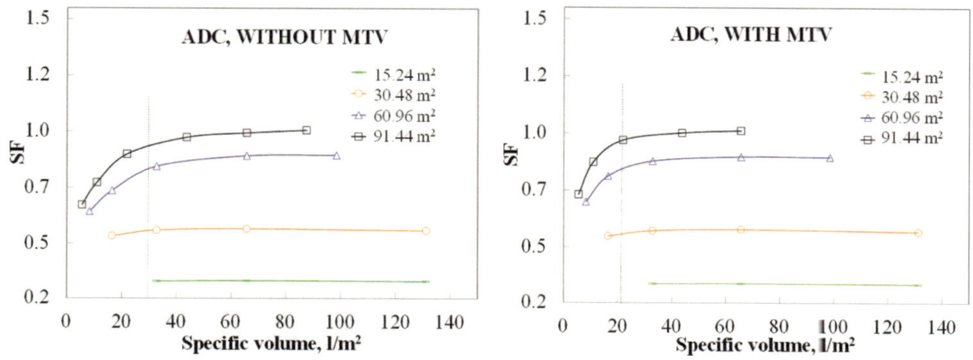

Figure 13.
Solar fraction against storage-specific volume for adsorption chiller.

(pumps, fans, and heater) as a function of the storage tank volume (ST) at 60.96 m^2 of the solar area. The calculated storage tank volume corresponds to 2.13 and 1.40 m^3 without and with MTV for absorption system and 1.64 and 1.34 m^3 for adsorption system, respectively. The use of storage volume higher than 3 m^3 has a soft effect on the Q_{HEATER} and W_{TOTAL} for absorption chiller with MTV and adsorption chiller.

Table 5 shows the values of the parameter showed in **Figures 16** and **17**. It was a little decrement of solar collector energy (from 1.77×10^7 to 1.61×10^7 kJ) with and without MTV for absorption chiller because the time of operation of the solar

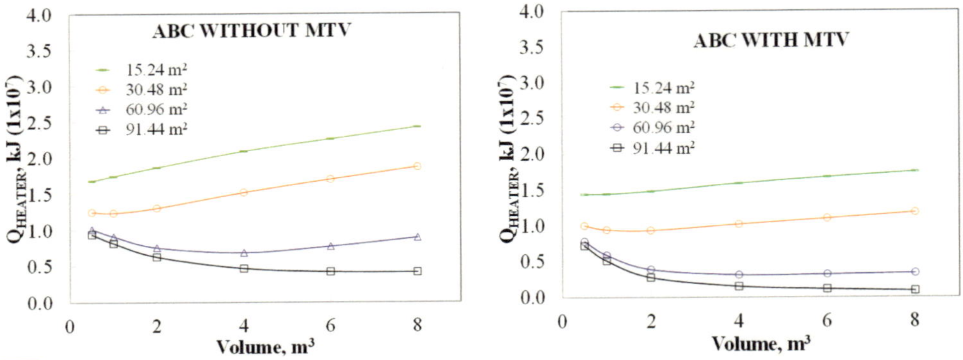

Figure 14.
Energy consumption of the heater against the storage volume for absorption chiller.

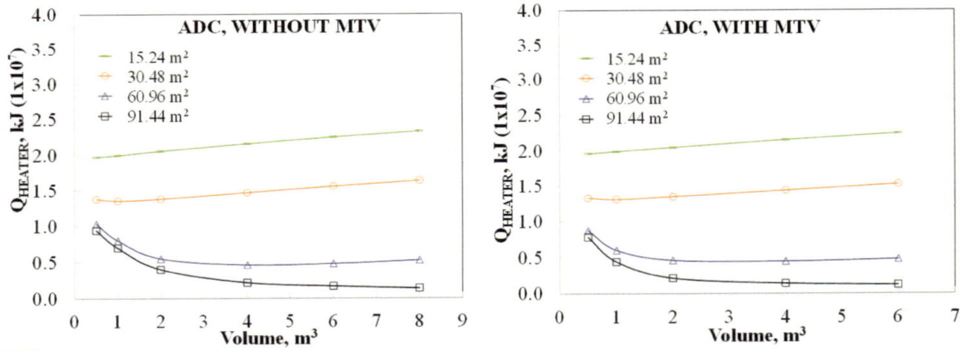

Figure 15.
Energy consumption of the heater against the storage volume for adsorption chiller.

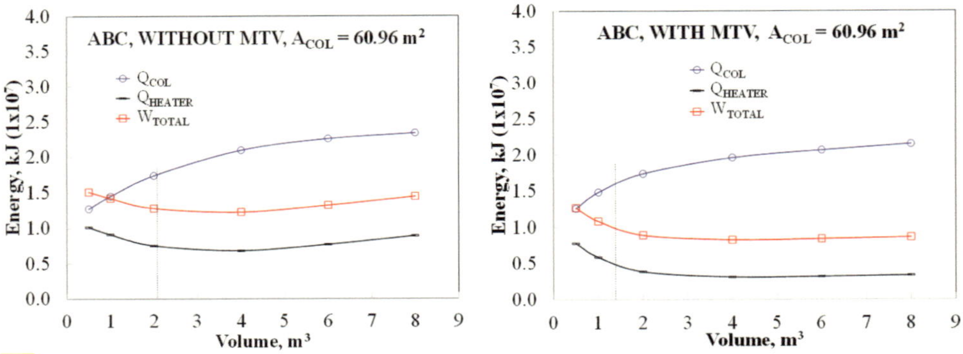

Figure 16.
The energy of the equipment against the storage volume for absorption chiller.

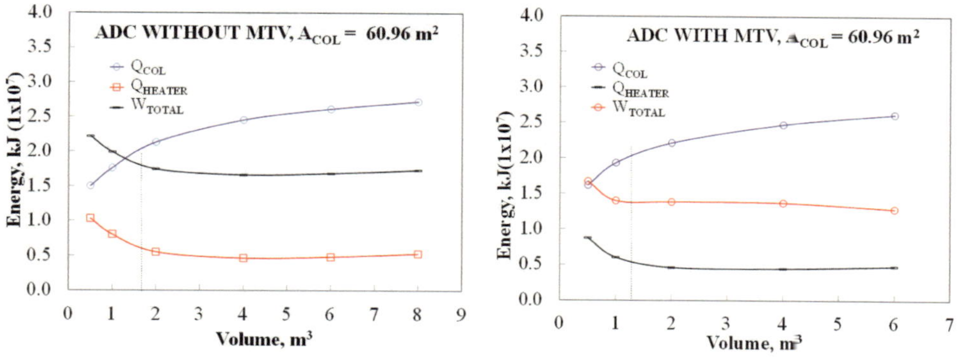

Figure 17.
The energy of the equipment against the storage volume for adsorption chiller.

	SV, l/m²	A_{COL}, m²	ST, m³	Q_{COL}, kJ (1×10^7)	Q_{HEATER}, kJ (1×10^7)	W_{TOTAL}, kJ (1×10^7)
ABS	35	60.96	2.13	1.77	0.74	1.27
ABS, MTV	23	60.96	1.40	1.61	0.48	0.95
ADS	27	60.96	1.64	2.02	0.62	1.69
ADS, MTV	22	60.96	1.34	2.05	0.52	1.44

Table 5.
Results of the selected conditions for cooling chillers.

collector was more using the MTV, then the temperature of the storage tank was higher and this reduces the efficiency of the solar collector, however it was a significant reduction on the Q_{HEATER} and W_{TOTAL} for both chillers, mainly for absorption chiller, this represents almost a 54.16 and 33.68%, respectively, while 19.23 and 17.36% of reduction was for adsorption chiller.

The reduction of the Q_{HEATER} is lower for adsorption than absorption chiller because the range of operation of the generator temperature is shorter for absorption than adsorption; however, W_{TOTAL} energy is higher for adsorption chiller than absorption chiller (1.27 kJ for absorption and 1.69 kJ for adsorption without MTV), because it has lower COP (around 0.53) than absorption chiller (around 0.70) and electric equipment have more time of operation.

5. Conclusion

This chapter presented a thermal analysis of the absorption and adsorption chillers in a dynamic condition for conditioning a building located in Temixco, Mexico, using TRNSYS and Microsoft Excel software from March to May. In this study, both chillers with and without MTV to increase the operation time using evacuated solar collectors were compared. The following conclusions are presented:

- The maximum solar collector energy was obtained with an angle of tilt of 7°.

- The COP has the higher values when using the input generator temperature. It selected a minimum temperature of working of 111 and 109°C for absorption chiller without and with MTV, and 75°C for adsorption chiller.

- The selection of 35 and 23 l/m^2 was a good choice for the configuration without and with MTV for absorption chiller, respectively, while 27 and 22 l/m^2 were selected for without and with MTV, respectively, for adsorption chiller.

- The working condition selected was 2.13 and 1.40 m^3 storage tank without and with MTV for absorption chiller and 1.64 and 1.34 m^3 without and with MTV for adsorption chiller using 60.96 m^3. The results showed a significant reduction of the Q_{HEATER} and W_{TOTAL} energy for both chillers, mainly for absorption chiller; this represented almost 54.16 and 33.68%, respectively, while 19.23 and 17.36% of reduction was for adsorption chiller, because adsorption has a lower COP (around 0.53) than absorption chiller (around 0.70) and electric equipment have more time of operation.

Nomenclature

Q	energy, kJ
m	mass flow rate, kg/s
Cp	heat capacity, kJ/kg C
T	temperature, °C
h	enthalpy, kJ/kg
COP	coefficient of performance
ST	store tank
SV	specific volume, l/m2
TW	total working
MTV	modulating tempering valve
Subscript	
CHW	chilled water
CW	cooling water
HW	hot water
AUX	Auxiliary system
HX	heat exchanger equipment
COL	solar collector

Author details

Jesús Cerezo Román[1*], Rosenberg Javier Romero Domínguez[1], Antonio Rodríguez Martínez[1] and Pedro Soto Parra[2]

1 Centro de Ingeniería y Ciencias Aplicadas, Universidad Autónoma del Estado de Morelos, Cuernavaca, Morelos, México

2 Instituto de Energías Renovables, Universidad Nacional Autónoma de México, Temixco, Morelos, México

*Address all correspondence to: jesus.cerezo@uaem.mx

References

[1] The future of cooling, opportunities for energy efficiency air conditioning, IEA. 2018. Available from: https://www.iea.org/publications/freepublications / publication/The_Future_of_Cooling.pdf [Accessed: December 11, 2018]

[2] Shirazi A, Taylor R, White S, Morrison GL. A systematic parametric study and feasibility assessment of solar-assisted single-effect, double-effect, and triple-effect absorption chillers for heating and cooling applications. Energy Conversion and Management. 2016;**114**:258-277. DOI: 10.1016/j.enconman.2016.01.070

[3] Kuczynska A, Szaflik W. Absorption and adsorption chillers applied to air conditioning system. Archives of Thermodynamics. 2010;**31**:77-94. DOI: 10.2478/v10173-010-0010-0

[4] Solar Energy Laboratory. TRNSYS 17 mathematical reference. In Available under the TRNSYS 17 Help Menu; Solar Energy Laboratory: Madison, WI, USA, 2017

[5] Solar Rating & Certification Corporation. Directory of SRCC Certified Solar Collector Ratings; Solar Rating & Certification Corporation: Washington, DC, USA; 2009

[6] Cerezo J, Romero RJ, Ibarra J, Rodríguez A, Montero G, Acuña A. Dynamic simulation of an absorption cooling system with different working mixtures. Energies. 2018;**12**:1-19. DOI: 10.3390/en11020259

[7] Khan M, Badar A, Talha T, Khan M, Butt F. Configuration based modeling and performance analysis of single effect solar absorption cooling system in TRNSYS. Energy Conversion and Management. 2018;**157**:351-363. DOI: 10.1016/j.enconman.2017.12.024

[8] Balghouthi M, Chahbani MH, Guizani A. Feasibility of solar absorption air conditioning in Tunisia. Building and Environment. 2008;**43**:1459-1470. DOI: 10.1016/j.buildenv.2007.08.003

[9] Calise F, d'Accadia MD, Vanoli L. Thermoeconomic optimization of solar heating and cooling systems. Energy Conversion and Management. 2011;**52**:1562-1573. DOI: 10.1016/j.enconman.2010.10.025

Section 3

Economic Prospects of Zero and Net Zero Homes

Chapter 6

Technical-Economic Research for Passive Buildings

Ruta Miniotaite

Abstract

The Energy Performance of Buildings Directive (EPBD) 61 requires all new public buildings to become near-zero-energy buildings by 2019 which will be extended to all new buildings by 2021. This concept involves sustainable, high-quality, valuable, healthy and durable construction. Foundation, walls and roofs are the most essential elements of a house. The type of foundation for a private house is selected considering many factors. The article examines technological and structural solutions for passive building foundation, walls and roofs. The technical-economic comparison of the main structures of a passive house revealed that it is cheaper to install an adequately designed concrete slab foundation than to build strip or pile foundation and the floor separately. Timber stud walls are the cheapest wall option for a passive house and 45–51% cheaper than other options. The comparison of roofs and ceilings showed that insulation of the ceiling is 25% more efficient than insulation of the roof. The comparison of the main envelope element efficiency by multiple-criteria evaluation methods showed that it is economically feasible to install concrete slab on ground foundation, stud walls with sheet cladding and a pitched roof with insulated ceiling.

Keywords: passive house, foundation, walls, roof, technological solutions, multiple-criteria evaluation

1. Introduction

A passive house is not a new method of construction. It differs from ordinary houses by good thermal insulation, high-quality windows and heat recovery ventilation system. All these features lead to the lower demand for thermal energy. The Energy Performance of Buildings Directive (EPBD) 61 requires all new public buildings to become near-zero-energy buildings by 2019 which will be extended to all new buildings by 2021 [1]. A passive house becomes a standard for energy-efficient buildings [2]. The problem in modelling a passive house occurs when investment into construction is estimated and the payback period for investment is calculated. The payback period depends on thermal energy price, which is difficult to forecast. Therefore a house of lower energy efficiency class is a less risky investment for an individual builder [3–5].

A passive house is the building with very low energy demand and uses only one-fourth or even less energy compared to the conventional buildings. Usually, passive houses do not have separate heating systems. Regenerative ventilation system is enough. The primary idea of such a house is to reduce the energy demand and at the same time maintain comfortable microclimate inside. The effectiveness of a passive house is based on the efficient thermal insulation and higher tightness of envelope components. A passive house is the concept (method) of construction applied in practice. A passive house is a standard widely used in constructing energy-saving buildings. A newly constructed passive house must save 80% of heat resources; otherwise it is not a passive house. The heating energy demand of a passive building is less than 15 k Wh/m^2 per year. However, a passive house is something more than just an energy-saving house [5–9]. This concept involves sustainable, high-quality, valuable, healthy and durable construction. Features of a passive house are the following: high insulation of envelope components, high-quality windows, good tightness of the building, regenerative ventilation system and elimination of thermal bridges. The recommended architectural solutions are simple forms, less angles in order to avoid the development of thermal bridges at the joints. The most effective form of the house is the one with the smallest area of external walls. For these reasons it is easier to build a house of bigger floor area meeting the passive house criteria because the area of external wall and thermal bridges is smaller than the useful floor area of the building. A passive house should have no basement; otherwise the basement must be well insulated. Besides, it is recommended to plan as many windows on the southern façade in order to use more natural solar energy. Windows must be made of special frames filled with double-chamber selective glass units. The site also has a significant effect on the energy demand by the building. Shadows from the neighbouring buildings must be considered when building a house in the district where tall buildings prevail. Water may be heated by electricity; however solar panels are recommended. Combination solar wind power generation units are recommended to produce electricity for lighting and regenerative systems as well as for household needs [10, 11]. This article examines technological and structural solutions for passive buildings' main structures.

2. Alternative solutions for passive house structures

2.1 Alternative solutions for foundation

Foundation is one of the most essential elements of a house. The type of foundation for a private house is selected considering many factors. The main factors are the type of the ground, groundwater level, frost line in the region, presence or absence of the basement, type of bearing walls, architectural decisions and financial resources. To choose the correct foundation type for the house, the builder must have the results of engineering and geological surveys, the final design of the building and calculations of loads.

2.1.1 Strip foundation

Strip is made of assembled concrete blocks or monolithic concrete. It is built under the bearing walls and partitions. This method requires land excavation, concrete element assembling and concrete pouring work. Construction of strip foundation is not cheap, but it is the most appropriate foundation for a house with a basement.

Monolithic strip foundation provides a more rigid framework, but the installation is longer than the foundation made of assembled concrete blocks. Monolithic strip foundation is recommended when the house is built on expanding soil.

Strip foundation is recommended when the house walls are made of heavier materials, for more than one-storey houses, and there is no need to build other types of foundations (e.g. pile foundation), which is necessary while building a house on weak, expanding or watery soil. Strip foundation is generally selected due to simple construction method, regardless the longer time required to build it.

2.1.2 Pile foundation

Pile foundation distributes the load of the building via pile cap and sides; therefore the stress propagates across the big volume of soil. Pile foundations do not sink much and have a high load-bearing capacity; thus they are suitable for buildings that are sensitive to subsidence.

As the piles are driven deep into the ground, it is impossible to fully insulate the entire foundation. Thermal bridges occur at the pole and grade beam joints, and they deteriorate the heat conservation capacity of the building.

2.1.3 Monolithic slab

A monolithic slab is a one-piece load-bearing foundation structure. Concrete is poured into special polystyrene foam forms that completely isolate the foundation slab from direct contact with the soil. It is the single type of foundation where the load-bearing monolithic slab has no contact with the soil and has the highest thermal resistance value. The thermal resistance value R of this foundation may be as high as 9.7 (R = 9.7 m^2 K/W) and higher. Thus, thermal bridges, frost and foundation deformations are avoided. The monolithic slab bears the load of the building across the entire plane rather than individual segments. The monolithic slab has from 3 to 20 times bigger supporting area than conventional foundations. For this reason it is less susceptible to movement and is firm and stable. All traditional foundation structures create thermal bridges because there are no structural possibilities to avoid them. Monolithic slab is the only exception where the entire concrete slab can be thermally insulated at 100%. A properly installed slab has no thermal bridges, and the main advantage of this foundation is high thermal resistance and tightness.

Estimate calculations of strip, pile and monolithic slab foundations are done. A private two-storey house is selected for the calculation. The calculated foundation area is 110 m^2. Labour, materials, machinery and total costs for foundation installation are calculated.

2.2 Alternative solutions for walls

2.2.1 Timber stud wall with sheet cladding

Double-stud wall structure significantly decreases the formation of thermal bridges and enables to diminish the weight of the entire structure. Thermal insulation layer is installed inside the double stud without an additional frame. The insulation layer thickness may vary depending on the required heat transfer factor U value. Thermal insulation made of PAROC WAS 25t sheets simultaneously serves as a wind barrier. This layer is fixed onto the studs. It is one of the best structures for a passive house (**Figure 1a**). An auxiliary frame on the internal side is required to install a tight vapour insulation, which in this system also serves as an air barrier. The vapour insulation is installed between the auxiliary internal frame and the thermal insulation layer in the middle. For this reason the engineering systems installed in the wall structure will not damage the tightness of the insulation layer. Four hundred and twenty millimetres of thick thermal insulation layer enables to achieve the U value ≤ 0.09 W/(m^2 K) [12].

a b c

Figure 1.
Structural solutions for walls: (a) timber stud wall with sheet cladding, (b) glued laminated timber I-joist stud wall with brick finishing and (c) plastered silicate block wall.

2.2.2 Glued laminated timber I-joist stud wall with brick finishing

Glued laminated timber (glulam) I-beam frame significantly reduces the impact of thermal bridges on the structure compared to the ordinary stud wall (**Figure 1b**). Thermal insulation made of PAROC WAS 25t sheets simultaneously serves as a wind barrier. This layer is fixed onto the studs. U value is ≤ 0.09 W/(m^2 K) when the thickness of thermal insulation layer is 420 mm [12].

2.2.3 Plastered silicate block wall

For the thermal insulation of plastered façades, the best option is rock wool sheets with vertically oriented fibre structure or special boards for plastered façades. These boards are fixed with adhesives onto the brick or concrete wall. The plates are covered by reinforcing layer and finishing plaster coat. Mineral or silicate decorative plasters should be used for such façades because they have better permeability for water vapour, i.e. create breathing walls (**Figure 1c**).

2.3 Roof and floor alternative solutions

The building of a passive house also involves the choice between a flat and pitched roof. If a pitched roof is selected, then there is a choice between thermal insulation of the roof and insulation of the ceiling with a cold attic. As in the case of walls, additional weight and thickness of roof insulation material must be considered. For the insulation of the entire pitched roof, the rafter height can reach up to 400–500 mm. Composite glulam rafter goes across the entire width of the thermal insulation layer. For the insulation of ceiling, less thermal insulation material is required, and the beam height may be lower; however the spans between beams must be narrower. Ceiling can be insulated not only with sheets but also with bulk insulating materials. In flat roofs the load-bearing structure is made of reinforced concrete slabs. Thermal insulation is installed in several layers with ventilation channels.

2.3.1 Thermal insulation of ceiling with an attic

Thermal insulation is made of three or more layers without any gaps between sheets and by overlapping the joints of the previous row (**Figure 2a** and **b**).

DOI: http://dx.doi.org/10.5772/intechopen.85992

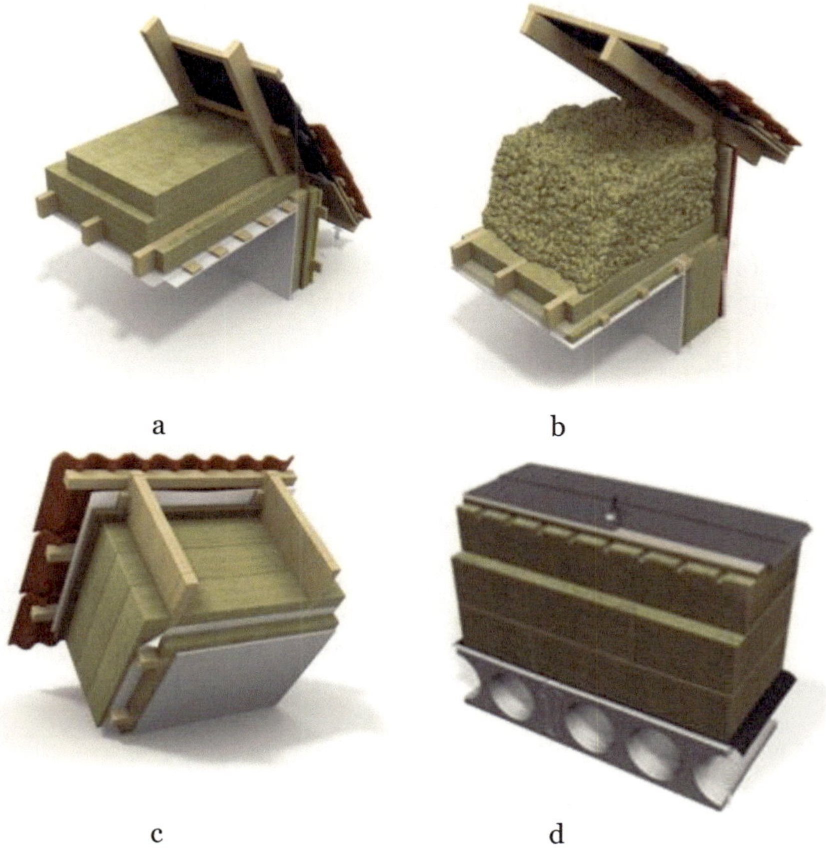

Figure 2.
Structural solutions for roofs: (a and b) thermal insulation of ceiling with an attic, (c) thermal insulation of a pitched roof and (d) thermal insulation of a flat roof.

2.3.2 Thermal insulation of a pitched roof

Rafters with bigger cross-section or glulam beams are used for pitched roof structure of a passive house. An auxiliary frame on the internal side is required to install an auxiliary thermal insulation layer and a tight vapour barrier, which in this system also serves as an air barrier. The vapour barrier is fixed to the bottom of the rafter. With the total thermal insulation layer of 550 mm, the *U* factor value is ≤ 0.07 W/(m^2 K) [12] (**Figure 2c**).

2.3.3 Thermal insulation of a flat roof

The roof of a passive house must be made of at least three layers. We recommend a ventilated PAROC Air structural solution where the insulation part of the intermediate layer has ventilation channels. *U* value is ≤ 0.07 W/(m^2 K) when the thickness of thermal insulation layer is 550 mm (**Figure 2d**).

3. Methods

Design solutions in construction can be evaluated by using different methods. According to the number of criteria, they are divided into single-criteria and multiple-criteria evaluations. In single-criteria evaluation of construction design solutions, construction costs of implementing alternative design solutions are calculated. The most effective alternative is selected according to this criterion [13]. However, construction projects and processes are multifaceted, complex and complicated. For this reason they are analysed by means of multiple-criteria decision-making. Construction projects and processes are multifaceted, complex and complicated.

The following criteria were used in our case:

- Technical: structural reliability of the system, noise level, universality of the building and degree of construction process mechanization
- Legal: environmental issues and occupational safety
- Economic: building site size, construction process duration, expenses and productivity
- Social: forms of labour organizations and motivation level

In this paper two evaluation methods were chosen: cost-benefit analysis and complex proportional assessment (COPRAS) method. Structures of energy efficiency class A house and passive house are compared. The main criteria for the evaluation of building structures are:

- Economic (construction price, length)
- Technological (complexity of technology, quality assurance level)
- Thermal parameters of the structures (thermal resistance, thermal bridges)

3.1 Cost-benefit analysis

In this analysis qualitative characteristics are measured by an expertise method while giving the scores in the grading scale 1–10. Ten is the best score. The criteria are not equally important; therefore the importance of one criterion with respect to another criterion is considered. All calculations and data are presented in a matrix table. The alternative with the highest cost-benefit value N is selected. This method enables to compare the analysed alternative in a simple and fast manner [13].

The first step is to select criteria for selected options. Criteria of economic, technological and thermal parameters were selected in order to evaluate different structures. Economic criteria include the cost of material, labour costs, cost of machinery and construction time. Technological criteria include the complexity of construction technology and quality assurance. Thermal parameters of the structures include the thermal resistance of a structure and elimination of thermal bridges.

The second step is to measure the weight (importance) of different criteria. In this paper the best options of technical-economic solutions for a passive house and class A house are analysed; therefore the biggest significance is given to construction price and thermal parameters of the structures.

The third step is to find the utility values of different options and evaluate them by scoring from 1 to 10. Explanation of utility values (from 1 to 10):

Score 10 for the cost of materials, labour costs and cost of machinery means that the amount of money spent to build the structure is the lowest. Other scores show relatively the difference between the prices of the analysed options.

Score 10 for construction time means that the structure was built in the shortest time compared to other analysed options. Other scores show relatively the difference between the construction time of the analysed options. Score 10 for the complexity of construction technology means that the technology of that option is the easiest and the most accessible compared to other options. Quality assurance shows how easy it is (more experience available) to install the structure of the analysed option. Score 10 means that it is easy, and score 1 shows that it is almost impossible.

Score 10 for the elimination of thermal bridges means that thermal bridges are minimized in that option. Other evaluations show relatively how effectively thermal bridges are eliminated in the relevant structure. Thermal resistance factor shows how well heat is protected in that structure option compared to other options. The option with the best thermal resistance factor is scored 10, and other options are scored according to their ability to retain heat.

The fourth step is the calculation of efficiency values taking into consideration the criterion importance. Utility values of different options are multiplied by the criterion importance:

$$b_{ij} = q_i \cdot x_{ij}, i = \overline{1, m};\ i = \overline{1, n} \tag{1}$$

where x_{ij} is the criterion i value for solution j, m is the number of criteria, n is the number of compared options and q_i is criteria significance.

In the fifth step, efficiency values of different criteria for all options are summed up:

$$N_j = \sum_{i=1}^{m} b_{ij}, i = \overline{1, m};\ i = \overline{1, n} \tag{2}$$

where N_j is the efficiency value of the solution option.

The best option is selected in the sixth step. The best variant is found after comparing the efficiency values among the options. The option with the highest efficiency value is the best solution.

3.2 COPRAS method

Goal setting, design and construction processes together with the final construction product and the subsequent operation process form one entity. When separate processes (solutions) of a project improve or deteriorate, the viability of the remaining solutions as well as stakeholders' satisfaction level changes accordingly. Therefore, a precise evaluation and calculation of the effect of all changes on the eventual outcome are important. To this end a complex proportional assessment [13, 14] method is used. Meanwhile, the priority and significance of analysed options directly and proportionately depend on the system of adequately describing criteria, criteria values and significant values. The criteria system is selected, and criteria values as well as initial significance are calculated by experts. Stakeholders (contractor, users, etc.) may modify all this information according to their goals and present circumstances. Therefore, evaluation of the options presents in detail the initial data provided jointly by the experts and stakeholders. The priority and significance of analysed alternatives are calculated in four steps [14, 15]:

1. A normalized decision matrix D is drawn. The goal of this step is to obtain dimensionless (normalized) estimated values from the compared criteria. When normalized estimated values are known, all indicators measured in different units can be compared. The calculation is done by using the formula

$$d_{ij} = \frac{x_{ij} \cdot q_i}{\sum_{j=1}^{n} x_{ij}}, \ i = \overline{1,m}; \ i = \overline{1,n} \tag{3}$$

where x_{ij} is the criterion i value for solution j, m is the number of criteria, n is the number of compared options and q_i is the criteria importance.

The sum of normalized estimated values j_i d of each criterion x_i is always equal to the importance q_i of that criterion:

$$q_i = \sum_{j=1}^{n} d_{ij}, \ i = \overline{1,m}; \ i = \overline{1,n} \tag{4}$$

The analysed criterion importance value q_i is distributed proportionally to all alternatives a_j with respect to their values x_{ij}.

2. The sums of normalized estimated minimizing (the lower value is better, e.g. price) criteria S_{-j} and maximizing (the higher value is better, e.g. quality) criteria S_{+j} that describe the alternative j are best calculated from the equation

$$S_{+j} = \sum_{i=1}^{m} d_{+ij}; \ S_{-j} = \sum_{i=1}^{m} d_{-ij}, \ i = \overline{1,m}; \ i = \overline{1,n} \tag{5}$$

In this case S_{+j} and S_{-j} values express the level of achieving the goals of the stakeholders of each alternative project. In any case the sums of "pluses" and "minuses" of all alternative projects are always equal to the sums of all maximizing and minimizing criteria values.

3. The relative significance (effectiveness) of compared options is found from their positive ("pluses" of the project) and negative ("minuses" of the project) characteristics. The relative significance Q_j of each variant a_j is found from the formula

$$Q_j = S_{+j} + \frac{S_{-\min} \cdot \sum_{j=1}^{n} S_{-j}}{S_{-j} \sum_{j=1}^{n} \frac{S_{-\min}}{S_{-j}}}, j = \overline{1,n} \tag{6}$$

4. The evaluated options are prioritized. The higher is the Q_j value, the more effective the option is. The method allows to easily evaluate and then select the most feasible solution with a clear physical view of the process. A generalized (reduced) criterion Q_j directly and proportionally depends on the relative influence of the compared criteria values x_{ij} and importance q_i for the final result.

4. Results and discussions

The comparison of different foundation options revealed that strip foundation is the most feasible for houses with basements built on very good soil conditions. Drilled piles are currently the most common foundation type due to economy and fast installation. However, the biggest disadvantage of this foundation for a passive house is the unavoidable thermal bridge at the pole and grade beam joints. Thermal insulation of these spots is almost impossible, and it is a doomed thermal bridge that should be avoided in a passive house.

A monolithic slab is the most appropriate foundation for passive houses due to its closed insulation circuit. Another advantage is suitability for different soil types. Besides, water supply and sewerage systems, power cables, heating system and subfloor, or sometimes even the normal floor, are installed together with the pile foundation. To this end very precise drawings of the house are required with all engineering and utility systems planned in advance. No significant changes of the house design are possible at later stages of construction. This disadvantage is eliminated by good planning and deliberations about the future use of the house. A monolithic slab becomes the most economic variant after the price of ground floor installation is added to the strip or pile foundations. It is 76% cheaper than pile foundation with ground floor installation and twice cheaper than strip foundation with ground floor installation.

4.1 Comparison of the efficiency among foundation options

The analysed options of a passive house foundation structures are as follows: F1P, strip foundation together with the first storey floor; F2P, pile foundation together with the first storey floor; and F3P, concrete slab floor.

The analysed options of energy efficiency class A house foundation structures are as follows: F1, strip foundation with first storey floor; F2, pile foundation with first storey floor; and F3, concrete slab floor. The obtained results are presented in **Figure 3**.

The comparison of utility values among the foundation options showed that concrete slab floor received the highest scores both in a passive house and in

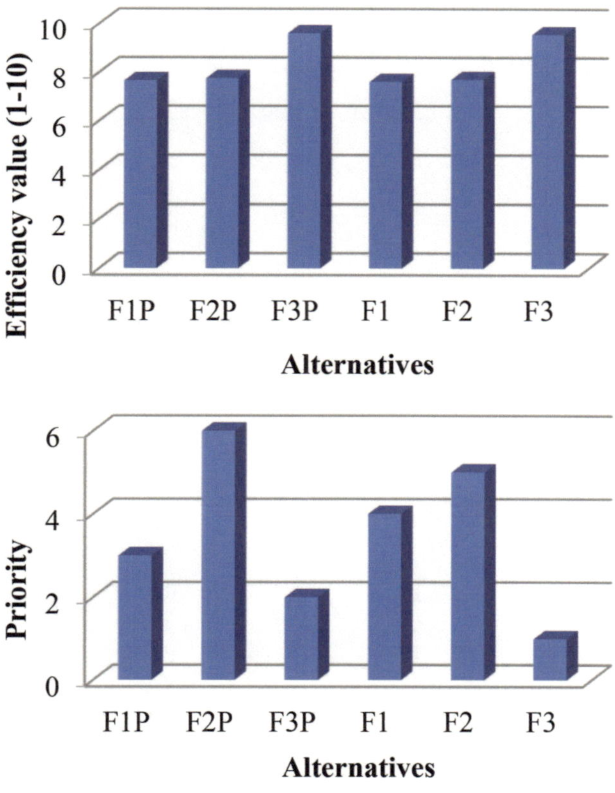

Figure 3.
Comparison of the efficiency among foundation options.

efficiency class A house. Efficiency values of concrete slab floor options were much higher than strip or pile foundations. The reason is much higher thermal parameters of the concrete slab floor than strip or pile foundation. The floor is installed together with the foundation slab and thus reduces the total price of the house. The analysis has shown that the best foundation option for energy-efficient houses is a concrete slab floor.

4.2 Comparison of the efficiency among wall options

The analysed options of a passive house wall structures are as follows: W1P, timber stud wall with sheet cladding; W2P, glued laminated timber I-joist stud wall with brick finishing; and W3P, plastered silicate block wall.

The analysed options of efficiency class A house foundation structures are as follows: W1, timber stud wall with sheet cladding; W2, glued laminated timber I-joist stud wall with brick finishing; and W3, plastered silicate block wall. The obtained results are presented in **Figure 4**.

The comparison of economic indicators for various options of passive house walls showed that timber stud walls are the best option. The cost of materials makes the major part in the price of plastered brick structures. Thermal insulation of such walls requires a thicker insulation layer in order to reach the passive house criteria for the walls. To improve the energy efficiency of such walls, the brickwork materials with better heat transfer factors should be chosen in order to have a thinner insulation layer. Timber stud walls require more labour, but 420 mm of the total thermal insulation layer is sufficient. The thermal insulation layer is installed between the load-bearing elements of the framework and auxiliary studs. The

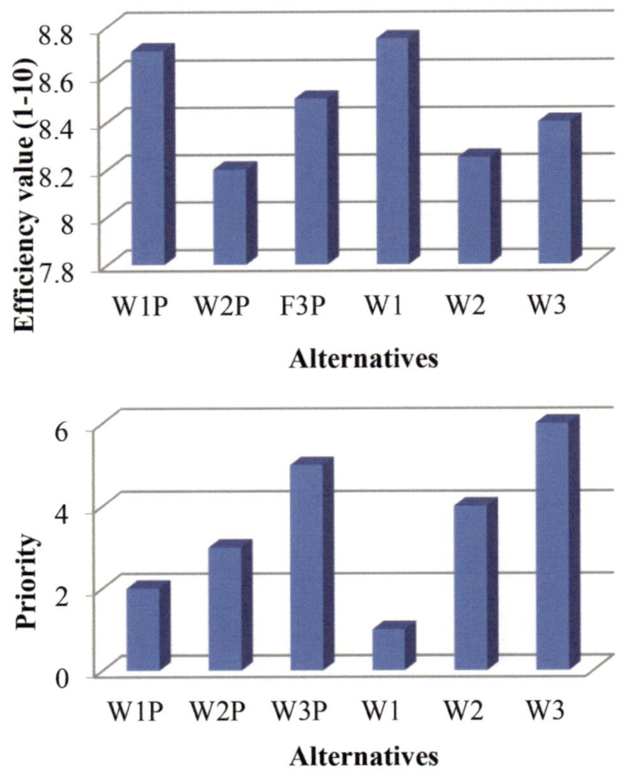

Figure 4.
Comparison of the efficiency among wall options.

load-bearing elements can be I-beams, box cross-sectional beams or glulam box beams. It should be noted that supervision of labour quality is vital for the building of such walls. Thermal insulation materials are installed in several layers. The tightness of air and vapour barrier must be ensured. According to the obtained data, the biggest price difference of alternative options is caused by the selected façade finishing. The brick finishing façade on stud walls increases the price of such walls significantly compared to sheet cladding finish. The sharp rise in plastered brick wall price is caused by the selected finishing plaster on the exterior. Timber stud walls with sheet cladding are 45% more efficient than silicate block walls and 51% more efficient than glulam I-joist stud walls with brick finishing.

4.3 Comparison of the efficiency among roof options

The analysed options of a passive house wall structures are as follows: R1P, insulation of a pitched roof ceiling with a cold attic; R2P, insulation of a pitched roof; and R3P, insulation of a flat roof concrete slab.

The analysed options of efficiency class A house foundation structures are as follows: R1, insulation of a pitched roof ceiling with a cold attic; R2, insulation of a pitched roof; and R3, insulation of a flat roof concrete slab. The obtained results are presented in **Figure 5**.

The comparison of flat and pitched roofs of the passive house with a cold and thermally insulated roof showed that a timber frame roof with thermally insulated ceiling and a cold attic is the best option for a single-family house. The labour cost indicator for a flat roof is economically the best compared to other alternatives;

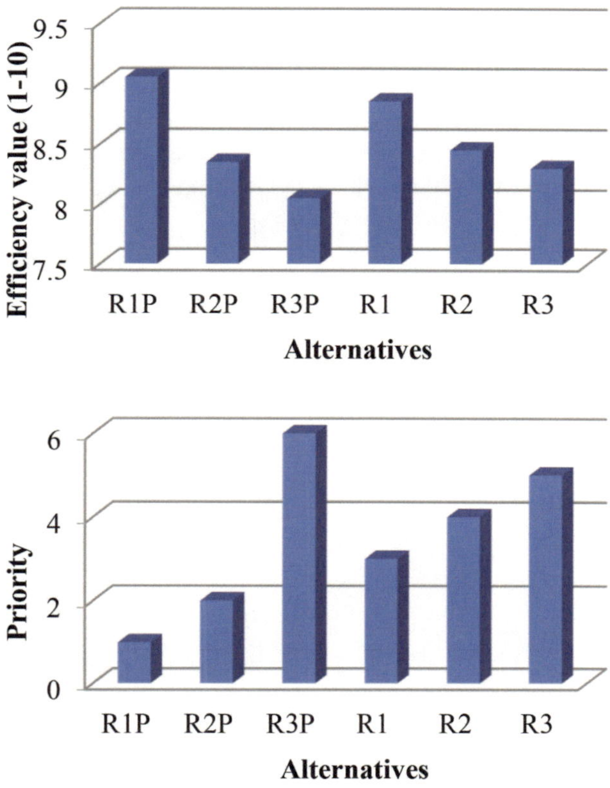

Figure 5.
Comparison of the efficiency among roof options.

however the price of the supporting structure and roof coat can be several times higher than in other options. Therefore, the total price of such a roof increases significantly. Insulation of the entire pitched roof requires much more insulation materials, and the height of the main rafters must be increased. Therefore, the total labour costs of such a roof increase significantly. The best economic effect is achieved by leaving the pitched roof uninsulated and installing the thermal insulation layer on the ceiling. The article analyses two different insulation options: using only insulation sheets and combining insulation sheets with bulky insulation material. Although the price of bulky insulation materials is lower, special blowing equipment is required. The combined insulation layer is also thicker. Attic insulation by means of rock wool sheets only is by 2% more economic than insulation with bulky foam and 25% cheaper than installing thermal insulation between rafters. A flat roof with concrete slab ceiling is several times more expensive than other options; therefore it is not recommended.

5. Conclusions

According to the passive house standard, an energy-efficient house is a building where energy is saved by architectural, structural and technological solutions. To meet the passive house standard requirements, the house must be tight, use renewable energy and employ various energy-saving methods and the planning done with respect to orientation.

The technical-economic comparison of the main structures of a passive house revealed that it is cheaper to install an adequately designed concrete slab foundation than to build strip or pile foundation and the floor separately.

Timber stud walls are the cheapest wall option for a passive house and 45–51% cheaper than other options. The comparison of roofs and ceilings showed that insulation of the ceiling is 25% more efficient than insulation of the roof.

The comparison of the main envelope element efficiency by multiple-criteria evaluation methods showed that it is economically feasible to install concrete slab on ground foundation, stud walls with sheet cladding and a pitched roof with insulated ceiling.

Conflict of interest

The author declares no conflict of interest.

Author details

Ruta Miniotaite
Kaunas University of Technology, Kaunas, Lithuania

*Address all correspondence to: rutaminiot@gmail.com

References

[1] The Energy Performance of Buildings Directive [Internet]. Available from: http://eur-lex.europa.eu/LexUriServ/LexUriServ.do?uri=OJ:L:2010:153:0013:0035:EN:PDF [Accessed: 20 December 2018]

[2] Passive House Institute. Passive House Requirements [Internet]. 2017. Available from: https://passivehouse.com/02_informations/02_passive-house-requirements/02_passive-house-requirements.htm [Accessed: 20 November 2018]

[3] Sun F. Critical review of EU Passive House Development [Internet]. Available from: http://www.irbnet.de/daten/iconda/CIB_DC26568.pdf [Accessed: 22 November 2018]

[4] Audenaert A, De Cleyn SH. Cost benefit analysis of passive houses and low energy houses compared to standard house. International Journal of Energy. 2010;**3**(4):46-53

[5] The European Near Zero Energy Buildings Project: The Greatest Energy and Green Building Project Ever [Internet]. Available from: http://www.house-energy.com/NZEB/UE-ZNEB.html [Accessed: 20 December 2018]

[6] Troung H, Garvie AM. Chifley Passive House: A Case Study in Energy Efficiency and Comfort. [Internet]. 2017. Available from: https://www.sciencedirect.com/science/article/pii/S1876610217334744 [Accessed: 20 December 2018]

[7] European Institute for Inovacion. Promotion of Near Zero CO_2 Emission Buildings Due to Energy Use [Internet]. 2016. Available from: https://www.interregeurope.eu/zeroco2/ [Accessed: 20 December 2018]

[8] Kuzman MK, Graselj P, Ayrilmis N, Zbasnik-Senegacnik M. Comparison of passive house construction types using analytic hierarchy process. Energy and Buildings. 2013;**64**:258-263

[9] Adhikari RS, Aste N, Del Pero C, Manfren M. Net zero energy buildings: Expense or investment? Energy Procedia. 2012;**14**:1331-1336

[10] Mutis I, Hartman T. Advances in Informatics and Computing in Civil and Construction Engineering. Hamburg: Springer-Verlag GmbH; 2018; ISBN-10: 3030002195, ISBN-13: 9783030002190

[11] Torgal FP, Labrincha JA. Biotechnologies and Biomimetics for Civil Engineering. Hamburg: Springer-Verlag GmbH; 2014; ISBN-10: 3319092863, ISBN-13: 9783319092867

[12] Structural Solutions [Internet]. Available from: http://www.paroc.lt [Accessed: 20 December 2018]

[13] Ginevicius R, Podvezko V. A feasibility study of multicriteria methods' application to quantitative evaluation of social phenomena. Business: Theory and Practice. 2008; **9**(2):84-87

[14] Juodis A. Modeling and Optimization of Construction Processes. Kaunas: Technology Press; 2005

[15] Miniotaite R. Multi-criteria decision analysis of up-to-date construction technology. In: Komurlu R, Gurgun AP, Singh A, Yazdani S, editors. Interaction Between Theory and Practice in Civil Engineering and Construction. Fargo, ND: ISEC Press; 2016. pp. 1-6

Chapter 7

Economic Aspects of Building Energy Audit

Samuel I. Egwunatum and Ovie I. Akpokodje

Abstract

Within the practice of construction economics, cost-benefit audits are carried out by proprietary audits with the intention of reporting the adequacy of any action and decision taken, meeting planned objective of a project or by efficiency audit which requires a more concise and restrictive investigation (like energy optimization) for its reporting. The efficiency audit system is most appropriate for energy utilization and performance investigation since it seeks to compare actual level of energy uses as against planned targets. This economic audit system of building energy requires that information about the energy designs are collected by means of management information system (MIS), reestablishing the data collected, comparing potential energy financial parameters with actuals, establishing the possible causes of variance. This is often justified or validated by such techniques as budgeted energy cost variance analysis, present value depreciation method, profit variance analysis, and cash flow and financial criteria analysis.

Keywords: building energy, energy audit, thermal properties, cost variance analysis, profit variance analysis

1. Introduction

The basic idea that heat is a form of energy that flows from one point to another as a result of difference in temperatures though governed by the laws of thermodynamics suggests that it can take a pattern of distribution in space according to the medium it travels through. Going through the space by way of transfer and interacting with the bodies it comes in contact with is a function of the phase and provided that there is no such change in phase, the heat required to raise the temperature of a mass of a building element (m) by a temperature (T) is associated by Eq. (1):

$$Q = mct \; c = \text{being the specific heat capacity of the element} \tag{1}$$

where m = mass of building element; t = temperature of building element.

At the point of temperature, increase Eq. (2) heat is absorbed or evolved when there is a phase change such that:

$$Q = ml, l = \text{being the specific latent heat} \tag{2}$$

On the contrary, the absorption or evolution of heat causes heat loss to its surrounding, such that a reversal process of that nature makes heat to be loss at a certain rate with a typical building space. The rate of such heat loss brings about the cooling of the environment or space which is proportional to the excess temperature of the building space over the external temperature of its surroundings based on forced convection when the excess temperature is small.

2. Thermal properties in building

Within a building space, heat is distributed or transferred by three fundamental methods which names are:

2.1 Conduction

This method of heat distribution or transfer in building spaces is a resultant effect of kinetic energy transfers at molecular level in any of the three states of matter (solids, liquids and gases). Conduction method of heat distribution or transfer in buildings is naturally validated to flow in the direction of tapering temperature. Such conductive behavior of heat loss in buildings is attractively noticeable in opaque walls during the winter [1]. There are experimental agreements between thermal and electrical conduction pattern in solids. The earliest formal knowledge of heat conduction law through a medium of either solid, liquid or gas was idealized by Joseph Fourier who postulated the law of heat conduction transfer method in the early part of the nineteenth century [2]. We take a congruency of how heat is conveyed through building elements (materials) or its space to be analogous to heat conduction in the building. Firstly, Fourier stated based on experimental verification for a steady conduction that the rate of heat transfer in any medium (inclusive of building elements) by conduction Q is proportional to the temperature difference and the heat flow area impacted by the heat in such a way that the heat conduction rate Q is inversely proportional to the distance through which the conduction traveled [1, 2]. Clearly in mathematical expression Eq. (3), Fourier meant that

$$Q = -KA.\frac{dT}{dx} \tag{3}$$

with K = thermal conductivity of the materials (W/(m.k)); A = area through which heat flow occurred; and $\frac{dT}{dx}$= temperature gradient at any point in x been the space through which the heat flow.

The minus sing indicating a flow from higher point to a lower point. For a recourse to building walls made of brick, block, fiber, paneled steel with known wall thickness, conductivity of heat from outer skin to inner skin can be estimated by:

$$Q = KA.\frac{T_1 - T_2}{\Delta x} \tag{4}$$

with k = thermal conductivity (W/(m.k)); T1 = outer/higher temperature; T2 = inner/lower temperature; A = area through which conduction flowed; and Δx = thickness of materials in which the conduction occurred.

The property of heat causing the differential between outer and inner temperature value is occasioned by the material's resistance to heat which is obtained when the above expression is having the conductivity (k) related to the area (A) as Eq. (5):

$$Q = \frac{T_1 - T_2}{\Delta x/KA}, R = \frac{\Delta x}{KA} \; as \text{ unit thermal resistance} \tag{5}$$

In practice, the term is commonly referred to as *R*-value such that Eq. (6)

$$R_{th} = \frac{\Delta x}{k} = AR \tag{6}$$

Another pseudo form of measuring thermal conductance in materials is the *U*-value which is expressed as the reciprocal of the *R* value:

$$U = \frac{1}{R_{th}} \tag{7}$$

The possibility that a building wall is layered with different materials and different geometries suggests that Fourier laws cannot be restricted to single layer with uniform thermal resistance [1]. HVAC systems carries different insulating and piping materials which calls for Eq. (8) cylindrical examination of Fourier law of steady heat conduction in cylindrical coordinates [3]. Under such consideration,

$$Q = \frac{T_1 - T_2}{In(r_o/r_i)/2\pi kl} \tag{8}$$

with ri = outer radius; ro = inner radius; L = pipe length; and K = thermal conductivity.

As a general rule, to the effect of other geometries, a shape factor (*S*) is introduced Eq. (9) to accommodate any derived shape for the measurement of heat loss of pipes in buried walls conveying hot fluid as:

$$Q = KS\Delta T = KS\,(T_1 - T_n) \tag{9}$$

Usually shape factors are derivatives of components meant in design to restrain losses which are either isothermal cylinder Eq. (10) with *S*-value of

$$S = \frac{2\pi L}{Cosh^{-1}(D/r)} \tag{10}$$

where $L \gg r$; and D = buried depth in semi-infinite medium Eqs. (11) and (12).

$$S = \frac{2\pi L}{In\ (2D/r)} \quad \begin{matrix} L \gg r \\ D < r \end{matrix} \tag{11}$$

$$S = \frac{2\pi L}{In\left[\frac{L}{r}\left(1 - \frac{In\ \left(\frac{L}{2d}\right)}{In\left(\frac{L}{r}\right)}\right)\right]} \quad \begin{matrix} D \gg r \\ L \gg r \end{matrix} \tag{12}$$

There can also be conduction between two isothermal cylinder Eq. (13) buried in infinite medium with *S*-value of

$$S = \frac{2\pi L}{Cosh^{-1}\frac{D^2 - r_1^2 - r_2^2}{2r_1 r_2}} \quad \begin{matrix} L \gg r_1 r_2 \\ L \gg D \end{matrix} \tag{13}$$

It can also take the form of conduction through two composite rectangular plane sections with edge section of two adjoining walls having combined k-value and inner/outer surface uniform temperatures *S*-value of Eq. (14)

$$s = \frac{al}{\Delta x} + \frac{bl}{\Delta x} + 0.54 \tag{14}$$

2.2 Convention

Unlike heat conduction where the transference of heat is through a body (solid) without visible motion of any part of the body to the naked eyes, convention is a method of heat transfer in fluid by the movement of the fluid itself [4]. Heat convention primarily takes two forms:

a. **Natural (or free)** heat convention is when the motion of the fluid is due solely to the presence of the hot body in it giving rise to temperature with a resultant mediums' density gradient causing the fluid to move under the control or restriction of gravity.

b. **Forced** heat convention is a process where heat is transferred with relative motion between the hot body and the fluid maintained by some external agency such as draught, making the relative velocity to contribute to the gravity current negligibly.

Free convention particularly results in density differences in the fluid, occasioned by contact with the surface originating the heat transfer [5]. Free conventions are evident in gentle air circulation in rooms due to solar-warmed windows or walls. On the other hand, forced convention occurs from the effect of an external force. Beside gravity to the problem, fluid moves past a warmer or cooler surface in obedience to Newton's laws of cooling. Under the two considerations, fluid velocities in free convention are considerably lower than fluid velocities in forced convention. Efficiency of heat transferred is a direct consequence of greater mechanical energy consumed in forced flow situation [1, 2]. Forced convention is seen applicable in the heat transfer process from heating and cooling coils. Convention is majorly responsible for cooling of buildings making it a common mode of heat transfers in buildings (**Tables 1–3**).

As a fall out of Newton's law of cooling which simply states that the rate at which heat is transferred by convention is proportional to the temperature difference and the heat transfer area. Mathematically, Newton by the law Eq. (15) is expressed as:

$$Q = h_{con}\, A\, (T_s - T_f) = h_{con}\, A\, (\Delta T) \tag{15}$$

with h_{con} = convention coefficient ($W/m^2.k$)); A = surface area through which convention occurs; Ts = surface temperature; and Tf = fluid temperature, away from wall.

Material	Specific Mass (kgm³)	Thermal Conductivity W/m²K)
brick	1600-1900	0.6-0.7
marble	2563	2.2
Gravel Concrete	2300-2500	1.7
Light Concrete	1600-1900	0.7-0.9
Glass	2500	0.8
Gypsum	1300	0.5
Hardwood	800	0.17
Softwood	550	0.14
Plywood	700	0.17
Floor Tiles	2000	1.5
Asphalt	2100	0.7
Linoleum	1200	0.17

Table 1a.
Thermal conductivity of common building materials (a, b, c).

Properties	Concrete	EPS[&]	Glass Fibre (Reference)
Thickness, δ, mm (inch)	152.4 (6")	63.5 (2.5")	140 (5.5")
Thermal Conductivity, λ_{eff} (W/(m.K))	1.4	0.0332	0.039
Density, ρ (kg/m^3)	2,350	22.7	11.5
Specific Heat, Cp (J/(kg.K))	880	1,210 (ASHRAE)	840
Volumetric Heat Capacity, $\rho\, Cp$ (kJ/(m^3.K))	2,068	27.47	9.66
Thermal Diffusivity, $\alpha = \lambda_{eff}/(\rho\, Cp)$ (m^2/s)	6.77×10^{-7}	1.21×10^{-6}	4.04×10^{-6}
Characteristic Time Constant, $\tau = \delta^2/\alpha$ (hr)	9.53	0.93	1.35
Thermal Resistance, RSI = δ/λ_{eff} (m^2.K/W)	0.109	1.913	3.590
Total Thermal Resistance, R (ft^2 hr °F/BTU)	**22.3[#]**		**20.4[$]**

[&] Properties at 23°C
[$] value does not include the effect of thermal bridging due to 2" x 6" studs
[#] value does not include the effect of thermal bridging due to the plastic spanners

Table 1b.

	Thermal conductivity (W/m K)	Density (kg/m^3)	Specific heat (J/kg K)	Emissivity (–)
Clay tiles	0.9	1950	1000	0.86
EPS insulation	0.035	25	1470	0.55
Fir wood	0.12	550	2700	0.88
Reflective layer	/	/	/	0.01[a]

EPS: expanded polystyrene; /: not measured items.
[a]Value provided by the manufacturer.

Table 1c.

The best theoretical approach for analyzing heat convention is attained by parameters of dimensional analysis using mass, length, time and temperature as focal dimensions Eq. (16). For dynamically similar bodies, natural convention is measured by

$$\left(\frac{h_{con}.l}{\lambda T}\right) = f_1\left(\frac{l^3 g\alpha P^3 T}{\eta}\right).\, f_2\left(\frac{C\eta}{\lambda p}\right) \tag{16}$$

This expression contains three dimensionless groups which include the Nusselt number $\left(\frac{h_{con}.l}{\lambda T}\right)$, the Grashof or free convention number $\left(l^3 g\alpha P^3 T/\eta\right)$ and the Prandtl number $\left(\frac{C\eta}{\lambda p}\right)$ where f_1 and f_2 is assumed to be dependent on the shapes of the dynamic bodies involved which serves as equivalents of shape factors (s) in conduction mode. When the convention is not free, i.e., forced, the analyzing equation Eq. (16) takes the form of Eq. (17)

$$\left(\frac{h_{con}.l}{\lambda T}\right) = F_1\left(\frac{lvp}{\eta}\right).F_2\left(\frac{C\eta}{\lambda p}\right) \tag{17}$$

Since it is a forced convention, this expression omits the free component (Grashof) and introduces the Reynolds number to the expression $\left(\frac{lvp}{\eta}\right)$. With all the numbers having reference tables, it makes it easy for HVAC designers to measure.

Material	diffusivity [m^2/s]
Building materials	
Aluminum	97.5 x 10^-6
Iron	22.8 x 10^-6
Marble	1.2 x 10^-6
Ice	1.2 x 10^-6
Concrete	0.75 x 10^-6
Brick	0.52 x 10^-6
Heavy soil (dry)	0.52 x 10^-6
Glass	0.34 x 10^-6
Wood (oak)	0.13 x 10^-6
Thermal insulators	
Cork	0.038 x 10^-6
Glass wool	0.023 x 10^-6
Rock wool	0.022 x 10^-6
Expanded polystyrene	0.035 x 10^-6
Extruded polystyrene	0.026 x 10^-6
Polyuretane foam	0.023 x 10^-6
Phenolic foam	0.018 x 10^-6

Table 2.
Diffusivity of common building materials (a, b).

Metals		Gases		Building Materials		Other Materials	
Aluminum	235	Air (dry)	0.026	Asphalt	0.75	Cotton	0.04
Brass	109	Argon (gas)	0.016	Brick dense	1.31	Cotton wool	0.029
Copper	401	Carbon dioxide (gas)	0.0146	Brick, fire	0.47	Diamond	1000
Gold	314	Helium	0.15	Brick, insulating	0.15	Engine Oil	0.15
Iron	67	Hydrogen	0.18	Concrete	0.8	Graphite	168
Lead	35	Krypton (gas)	0.0088	Fiberglass	0.048	Ground or soil, dry area	0.5
Nickel	91	Methane (gas)	0.03	Polyurethane foam	0.024	Ground or soil, moist area	
Silver	428	Nitrogen (gas)	0.024	Rock wool	0.043	Polyethylene - low density	0.33
Sodium (liquid)	86	Steam, saturated	0.0184	White pine	0.11	Polypropylene, PP	0.1 - 0.22
Sodium (solid)	135	Xenon (gas)	0.0051	Window glass	1	Porcelain	1.5
Stainless steel	14			Wood, oak	0.17	Sulfur, crystal	0.2
Steel, Carbon 1%	43					Uranium dioxide	8.8
Thorium (metalic)	38					Water	0.58
Uranium (metalic)	27.6						
Zirconium	22.6						
Zirconium alloy (1% Nb)	18						

Table 2b.

A corresponding derivation for R value and thermal resistance exists for convention methods of transfer that serves for both forced and free conventions Eq. (18) as

$$R = \frac{1}{h_{con} A} \tag{18}$$

$$\text{With } a \text{ heat transfer function of } Q = \frac{\Delta T}{R} \tag{19}$$

Resistance to thermal effusion under convention with R_{th} value and the associated U-value are given by Eqs. (20) and (21)

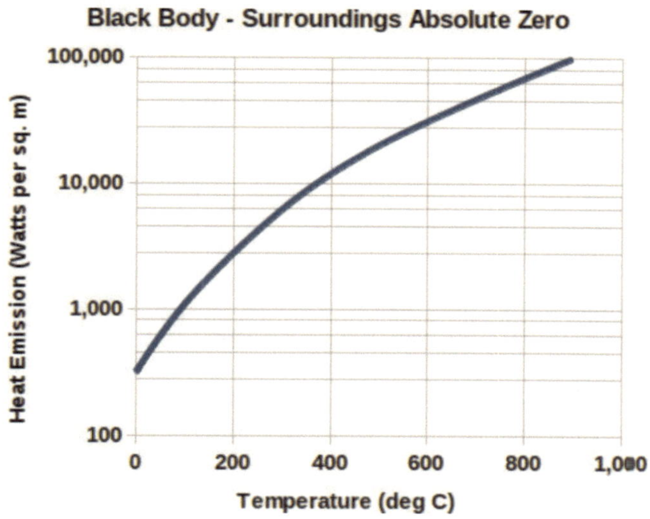

Table 3a.
Radiation on surfaces (a, b).

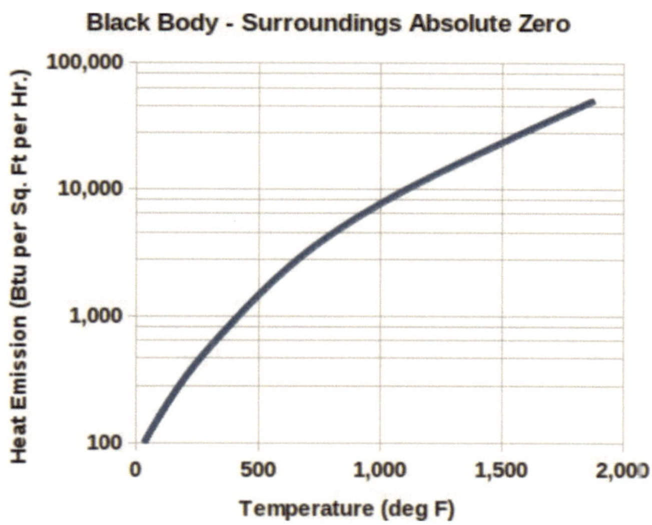

Table 3b.

$$R_{th} = \frac{1}{h_{con}} \tag{20}$$

$$U \equiv \frac{1}{R_{th}} = h_{con} \tag{21}$$

Heat flows outside of buildings have been a source of heating in the inside of buildings. It is well a good preemptive move to determine heat flows outside of buildings which naturally contributes to the heating in the entire building envelope [1, 3, 5]. Most external heat flows in buildings with forced convention flows are usually regarded as turbulent, but usually again take the form of laminar or turbulent flow when the convention currents are free [1–3]. Keeping units of measurements in SI units, there are experimental proofs that with air temperature between 19 and 21°C for interior walls and window surfaces having confluence, with exterior surfaces, laminar free convention of air from internal surfaces is given as Eq. (22)

$$h_{con} = 1.4232\left(\frac{\Delta TSin\beta}{L}\right)^{\frac{1}{4}} \tag{22}$$

wherein ΔT equates the temperature difference as $(T_s - T_f)$, with L as the length of the horizontal framing member on a vertical stud and titling surface in the direction of the buoyed-driven flow caused by the convention. β is the surface tilt angle with acute properties (30–90°) and the flow condition that $L^3\, \Delta T < 1.0$.

If $L^3\, \Delta T < 1$, a turbulent flow occurs for which turbulent free convention from a tilted surface in air gives Eq. (23)

$$h_{con} = 1.3131\ (\Delta TSin\beta)^{\frac{1}{3}} \tag{23}$$

For horizontal members' particularly horizontal pipes and cylindrical members in air, laminar free flow convention is estimated from Eq. (24)

$$h_{con} = 1.3131\ (\Delta T/D)^{\frac{1}{4}} \tag{24}$$

with D as the cylinder's outer diameter. Both turbulent and laminar flows have the same standards for test and measurement with those of tilted members and an adjustment for L with D. Notwithstanding, building elements with cylindrical components in air have their turbulent free flow convention computed from Eq. (25)

$$h_{con} = 1.2401\ (\Delta T)^{\frac{1}{3}} \tag{25}$$

However, structural members or surfaces say flat roots having complete 100% exposure horizontally to solar warming without recourse to solar angle have their laminar free flow convention coefficient estimated from Eq. (26)

$$h_{con} = 1.3203\ (\Delta T/L)^{\frac{1}{4}} \tag{26}$$

with L as the average uniform length of the horizontal surface. The flow condition for the above expression is also true for humid or cold surfaces in reversed contact with the sun as obtainable in the surface of a plane skylight in roof tops. Warm surfaces in direct exposure with solar light have their turbulent free convection coefficient computed from Eq. (27) turbulent flow as

$$h_{con} = 1.5214\ (\Delta T)^{\frac{1}{3}} \tag{27}$$

warmed surfaces not having direct surface exposures have their laminar convection coefficient reduced owing to stable stratification condition.

2.3 Radiation

This is the process whereby radiant heat energy is transferred from one point to another. It belongs to the class of electromagnetic spectrum between higher and radio waves with a range of wavelength between 740 and 0.3 mm approximately

[1, 5]. Heat radiations been electromagnetic in nature are originated by thermal movement of particles in matter. At temperatures higher than absolute zero, all matter sends out thermal radiation. The dynamical behavior or movements of particles results in charge acceleration that produces electromagnetic radiation. Thermal emitting bodies at any temperature consist of a wide range of frequencies [3]. Most radiating bodies have dominant frequencies which shift to higher frequencies as the temperature of the source increases. In most thermal radiation situations, the total amount of radiation for all frequencies increases sharply as the temperature rises at a rate of T^4, with '*T*' as the absolute temperature of the body [6, 7]. Consequently, the rate of the electromagnetic radiation emitted at a certain frequency is proportional to the amount of absorption that it would experience by the source [3, 6]. Estimation of heat transferred by radiation is called net radiative heat transfer, which is the heat transferred from one surface to another, been the heat leaving the first surface for the other and subtracting the heat arriving from the second surface.

For radiating black bodies, the radiation rate from Surface A to Surface B is given in Eq. (28):

$$Q_{1-2} = A_1 E_{b1} F_{1\to 2} - A_2 E_{b2} F_{2\to 1} \tag{28}$$

where A is surface area, E_{b1} is energy flux, $F_{1\to 2}$ is the view factor originating from surface 1 to surface 2.

Since the two surfaces exchange their heat loses, the reciprocity rule holds for the view factors as $A_1F_{1\to 2} = A_2F_{2\to 1}$ with heat flux emissive power of black body given as $E_b = \sigma T^4$ then $Q_{1-2} = A_1F_{1\to 2}\left(T_1^4 - T_2^4\right)$ where 'σ' is the Stefan-Boltzmann constant and T is absolute temperature. Should the value of Q give negative value; it suggests that net heat transfer is from surface 2 to surface 1.

As a departure for black bodies, two gray surfaces retaining an enclosure have their heat transfer rate given in Eq. (29):

$$Q = \frac{\sigma\left(T_1^4 - T_2^4\right)}{\frac{1-\epsilon_1}{A_1\epsilon_1} + \frac{1}{A_1F_{1\to 2}} + \frac{1-\epsilon_2}{A_2\epsilon_2}} \tag{29}$$

where ϵ_1 and ϵ_2 are the emissivity of the surfaces and any $\epsilon = \frac{E}{E_b} = \frac{\textit{actual emissive power}}{\textit{emissive power of black body}}$

Theoretically, the Stefan-Boltzmann law as stated in Eq. 28 above governs radiation emission of a blackbody (ideal radiator). Besides the emissivity of materials, other indices used for computing the rate of radiation heat transfer from surfaces includes absorptivity (α), transmissivity (τ) and reflectivity (ρ) [3, 5, 8]. All radiating surfaces have these three properties Eq. (30) related by the law of conservation of energy as:

$$\alpha + \tau + \rho = 1. \tag{30}$$

However, this relation depends on the nature of the wave-length been radiated which is verifiably true for single wavelengths and gray surfaces. For wavelengths whose range are over the three properties are calculated as same.

Temperature absorption questions through building elements with dark boundaries have been given extensive analysis in the works of [7] by integro-differential means in Eq. (31)

$$N_r \frac{d^2\theta(\tau)}{d\tau^2} = n^2(\tau)\theta^4(\tau) - 1/2\left[\beta(\tau)E_2(\tau) + \beta(\tau^\circ)E_2(\tau^\circ - \tau) + \int_0^{\tau^\circ} n^2(\tau^\circ)E_1(|\tau - \tau^\circ|)\theta(\tau^\circ)d\tau^\circ\right] \tag{31}$$

Noting the boundary condition to be

$$\theta(0) = \theta_2, \theta\tau^{\circ} = 1.0$$

whereas, [7] showed estimation of the absorbed heat with dimensionless temperature value which has equally been shown to be of value expressed in Eq. (32)

$$\theta(\tau^{\circ}) = G(\tau) = \frac{1}{2N_r}\int_0^{\tau^{\circ}} n^2(\tau^{\circ})\left\{-E_3(|\tau - \tau^{\circ}|) + E_3(\tau^{\circ}) + \frac{\tau}{\tau^{o}}[E_3(\tau^{\circ} - \tau) - E_3(\tau^{\circ})]\right\}\theta^4(\tau^{\circ})d\tau^{\circ} \tag{32}$$

Annotated by Eq. (33)

$$\begin{aligned} G(\tau) = \frac{1}{2N_r}\Bigg(&\beta(0)\left[-E_4(\tau) + \frac{\tau}{\tau^{o}}E_4(\tau^{\circ}) + \frac{1}{3}\left(1 - \frac{\tau}{\tau^{o}}\right)\right] \\ &+ \beta(\tau^{\circ})\left[\left(1 - \frac{\tau}{\tau^{o}}E_4(\tau^{\circ}) - E_4(\tau^{\circ} - \tau) + \frac{1}{3}\frac{\tau}{\tau^{o}}\right]\right. \\ &+ 2N_r\left\{\theta(0) + \frac{\tau}{\tau^{o}}\left[\theta(\tau^{\circ}) - \theta(0)1\right\}\right)\Bigg) \end{aligned} \tag{33}$$

In near real life situation, determination of temperature profiles in buildings have not been successful with closed-form solutions but with numerical methods Eq. (34) to obtain the total building heat flux through its elements as

$$\begin{aligned} q_t = &\frac{k_c}{L}(T_1 - T_2) + 2\sigma\left(T_2^4\left[E_3\tau^{\circ} + \frac{1}{\tau^{o}}E_4\tau^{\circ} - \frac{1}{3\tau^{o}}\right]\right. \\ &+ T_1^4\left[-\frac{1}{\tau^{o}}E_4\tau^{\circ} - \frac{1}{2} + \frac{1}{3\tau^{o}}\right] + \int_0^{\tau^{\circ}} n^2(\tau^{\circ})\left\{-E_2(\tau^{\circ} - \tau)\right. \\ &+ \frac{1}{\tau^{o}}[E_3(\tau^{\circ} - \tau) - E_3(\tau)]\}T^4(\tau)d\tau\Big) + \sigma T_1^4 - 2\sigma T_2^4 E_3(\tau^{\circ}) \\ &- 2\sigma\int_0^{\tau^{\circ}} n^2(\tau^{\circ})E_2(\tau^{\circ} - \tau)\left(T^4\tau\right)d\tau \end{aligned} \tag{34}$$

The radiative and conductive fluxes are closely outlined by the terms of Eq. 34 in such a way that the first two terms of Eq. 34 are suggestive of conductive flux, while the last three terms are radiative [8, 9]. Upon the combination of both integrals, Eq. 34 becomes Eq. (35)

$$\begin{aligned} q_t = &\frac{k_c}{L}(T_1 - T_2) + 2\sigma\left(T_2^4\left[\frac{1}{\tau^{o}}E_4\tau^{\circ} - \frac{1}{3\tau^{o}}\right] + T_1^4\left[-\frac{1}{\tau^{o}}E_4(\tau^{\circ}) + \frac{1}{3\tau^{o}}\right]\right. \\ &+ \int_0^{\tau^{\circ}} n^2(\tau^{\circ})\frac{1}{\tau^{o}}[E_3(\tau^{\circ} - \tau) - E_3(\tau)]\left(T^4\tau\right)d\tau\} \end{aligned} \tag{35}$$

with such algebraic treatment, Eq. 33 can as well be treated with integral calculus to give Eq. (36)

$$\frac{d\theta}{d\tau} = (1/2N_r)(\beta(0)[E_3(\tau) + E_4(\tau^o)/\tau^o - 1/3\tau^o]$$

$$+ \left(\beta(\tau^o)\left[\frac{-E_4(\tau^o)}{\tau}{}^o - E_3(\tau^o - \tau) + 1/3\tau^o\right] + (2N_r/\tau^o[\theta(\tau^o) - \theta(0)]\right. \quad (36)$$

$$+ \int_0^{\tau^o} n^2(\tau^o)\{E_2(|\tau - \tau^o|) + (1/\tau^o)[E_3(\tau^o - \tau) - E_3(\tau^o)]\}\theta^4(\tau)d\tau^o)$$

By securitizing Eq. 36, the steep behavior or temperature gradient of the absorption can be inferred from the dimensionless gradient (β) which satisfies the Schwartz inequality in Eq. (37)

$$\beta = \frac{d\left(\frac{\theta T_1 - T_2}{T_1 - T_2}\right)}{d\left(\frac{\tau}{\tau^o}\right)} \begin{vmatrix} \geq 1 \\ \tau = 0 \end{vmatrix} \quad (37)$$

In recent times, many simulation techniques have been developed to determine temperature profiles and heat fluxes in building, but many of which are interactive in nature.

This brings us to the absorptivity and emissivity of gray surfaces which under the Kirchoff's identify are equal, been

$$\alpha = \epsilon$$

Even for non-gray surfaces at a stipulated wavelength, drawing a clue from [5] experiment that the mean free path of a photon ($\frac{1}{E}$) passing through an object with thickness, L is related by the formula

$$\tau^o = \frac{L}{1/E} - EL \ll 1$$

This is premised on the notion that the radiant heat flux is not tempered by the material and provided the conductive radiative mechanisms are not acting with each other, the building element will experience a total heat flux Eq. (38) equal to

$$q_t = q_c + q_r \quad (38)$$

where q_t = total heat flux; q_r = radiative heat flux; q_c = conductive heat flux—by Fourier Law.

Such that planar elements with thickness L, having uniform properties and unidirectional steady state heat flow have their values computed from Eq. (39)

$$q_r = k_c\frac{(T_1 - T_2)}{L} \quad (39)$$

And radiant heat flux with two infinite parallel plates with temperatures at T_1, and having T_2 and emissivites ϵ_1 and ϵ_2 computed by Eq. (40)

$$q_r = \frac{\overline{r}(T_1^4 - T_2^4)}{\frac{1}{\epsilon_1} + \frac{1}{\epsilon_2} - 1} \quad (40)$$

with reference to Eq. (38), q_t and by substitution, reduces to Eqs. (41) and (42)

$$q_t = k_c \frac{(T_1 - T_2)}{L} + \frac{\sigma\left(T_1^4 - T_2^4\right)}{\frac{1}{\epsilon_1} + \frac{1}{\epsilon_2} - 1} \tag{41}$$

Since $\epsilon_1 = \epsilon_2 = 1$ for black plates, q_t becomes

$$q_t = k_c \frac{(T_1 - T_2)}{L} + \sigma\left(T_1^4 - T_2^4\right) \tag{42}$$

Besides [5] investigation for the optically thin limit case, the limiting case for the optical thickness limit was investigated by [10] for elements that are large compared to the mean free path of the photon causing the radiation, giving rise to conductive heat transfer process. By experiment, from [10]

$$\tau^o = \frac{L}{1/E} = EL \gg 1$$

So that radiant heat flux arising from radiant energy Eq. (43) becomes

$$q_t = -k_r \frac{dT}{dx} \tag{43}$$

And by combining the conductive and radiative heat transfers of the element, the total heat flux for the building element at a steady state for uni-directional heat flow as Eq. (44) becomes

$$q_t = k_c \frac{T_1 - T_2}{L} + \frac{4_n^2 \sigma\left(T_1^4 - T_2^4\right)}{3\alpha L} \tag{44}$$

where k_r = radiative conductivity of a gray medium $\cong \frac{16}{3} \frac{n^2 \sigma T^3}{\alpha}$

$$k_{app} = k_c + k_r$$

$$q_t = -k_{eff} \cdot \frac{dT}{dx}$$

With this totality conduction, apparent thermal conductivity is obtained by the relationship

$$k_{app} = q_t L / (T_1 - T_2)$$

With particular reference to the thickness (L) of the material, the conductivity through the optical element at constant temperatures of the plate as Eq. (45) becomes:

$$k_{app} = k_c + \frac{\sigma\left(T_1^4 - T_2^4\right)}{T_1 - T_2} L \tag{45}$$

Experiments have shown that the upper limit of the apparent thermal conductivity of the material greatly depends on the absorption coefficient of the material's thickness and extreme absorption coefficients. With these two conditions, k_{app} becomes Eq. (46)

$$k_{app} = k_c + \frac{4_n^2 \sigma Q\left(T_1^4 - T_2^4\right)}{3\alpha(T_1 - T_2)} \tag{46}$$

Discussing conductivity and heat radiation of building elements with respect to the element's thickness as it affects the apparent thermal properties of insulation has its credit due to [8]. The basic concept of [8] idea is that by the very nature of insulation, conduction and radiation does not occur and their individual heat fluxes are sums. Going by [8] theorem, at heat radiation equilibrium, radiant heat flux between two infinite parallel plates separated by a di-heat gray material or medium at temperatures T_1 and T_2 as given in Eq. (47)

$$q_r = \frac{\sigma\left(T_1^4 - T_2^4\right)}{1 + \left[\left(\frac{1}{\epsilon_1}\right) + \left(\frac{1}{\epsilon_2}\right) - 2\right]Q} \tag{47}$$

As stated in heat transfer literature, several methods exist for treating the effect of thickness on apparent thermal properties of insulation. But exact solution is found in the approach of [9] with the condition that $T^{o} \gg 1$; then

$$Q = \frac{4/3}{\tau^{o} + \gamma}$$

where $\gamma = 1.4209$.

Appropriate approximation to this problem is found in the *exponential-kernel* as

$$Q = \frac{4/3}{\tau^{o} + \frac{4}{3}}$$

While that of [8] is consistent with the *exponential-kernel* approximation, Rennex only introduced a factor in the approximation value of Eq. (48) by proposing that

$$Q = \frac{4/3}{\tau^{o} + \frac{4}{3}[factor]} \tag{48}$$

Factor = 1 + 0.0657 tanh (27°) while addressing the [9] Q-value estimation. A further theory on the effect of elements thickness on the apparent thermal conductivity with the assumption that radiative and conductive heat fluxes do not interact and premised on the computation that total heat flux is equal to the sum of the individual fluxes, [8] obtained the value of K_{app} by substitutions in Eq. (49)

$$K_{app} = k_c + \frac{4\sigma T_m^3}{\frac{2}{\epsilon} - 1 + \frac{3}{4}\tau^{o} + 0.0657} L$$

$$K_{app} = k_c + \frac{4\sigma T_m^3}{\frac{2}{\epsilon} - 1 + \frac{3}{4}\tau^{o}} L \tag{49}$$

Due to Rennex, Eq. (50) we have

$$K_{app} = \frac{4\sigma T_m^3}{\frac{2}{\epsilon} - 1 + \frac{3}{4}\tau^{o} + 0.0657 \tan\left(2\tau^{o}\right)} L \tag{50}$$

Provided $\epsilon_1 = \epsilon_2 = \epsilon$ and $\left(\frac{\left(T_1^4 - T_2^4\right)}{T_1 - T_2}\right) \cong 4T_m^3$

Thickness has its effects on the conductivity of building elements demands [9, 10]. On the whole, optical thickness of elements increases the materials thermal conductivity by asymptotic expansion which tends to a limiting value in such a way that apparent thermal resistance has a linear dependence on element's thickness as the element's thickness approaches infinity. In the same vein, apparent thermal resistivity of the element is equal to the apparent thermal resistance divided by the elements thickness [6, 9, 10].

3. Building energy analysis

Building energy analysis in the twenty first century is becoming a prerequisite for building comfort design and its universal acceptability as a practice is not unconnected to energy crises with the advent of increased global warming. Consequently building energy analysis is required right from the project conceptual/ scheme design stage to accommodate the integration of options and alternatives towards maximizing energy uses (see **Figures 1-5**) [1, 11]. This practice has some economic consideration with recourse to profit maximization of the project. In Europe, scheme designs must accommodate energy analysis report as a statutory requirement for building development by authorities with respect to various legislation backing it up [11]. Before now, building energy analysis is time consuming for experts, but in recent scheme development such arduous process have been responded to with Building Information Model (BIM) energy appraisals with the attendant aim of gaining project time and cost savings. Issues of time and cost in construction projects are economic indices of estimating project success and performance with varying degrees of their weights in a project. Most times, both variables are often contesting like lunar eclipse to avoid project escalations. The use of BIM in building energy analysis, strongly assist experts in avoiding complicated laborious calculations. Most common tool which software based in evaluating or analyzing building analysis are AUTODESK ECOTECT™, AUTODESK Green Building Studio™ Integrated Environment Solutions (IES) Virtual Environment and Revit™ [11].

To a large extent, energy uses and building performance largely depends on the envelop properties of the building since it attempts to balance energy transfer

Figure 1.
Residential building Revit model (kind permission from Jangalve et al. [3]).

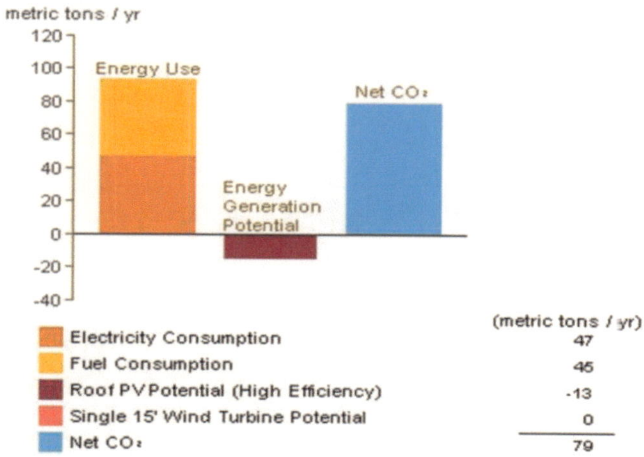

Figure 2.
Energy model of residential building.

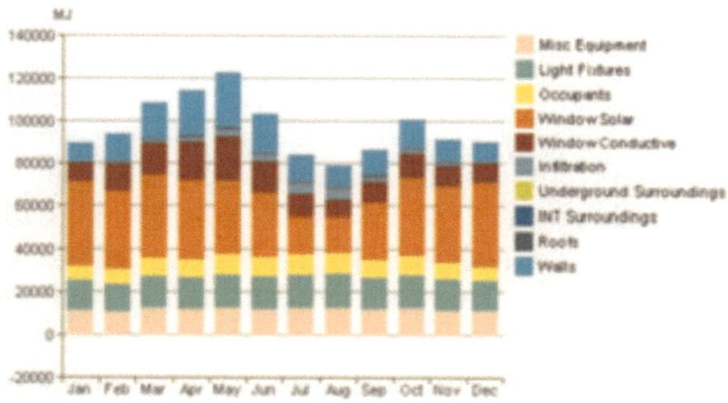

Figure 3.
Monthly cooling load.

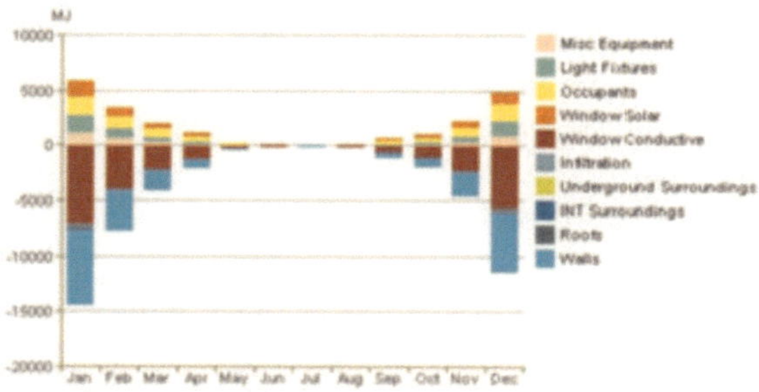

Figure 4.
Monthly heating load.

process between the internal and external environment. If properly handled after the scheme designs and documents are approved, energy analysis of the building envelope will give a clear cut direction for optimization and systems sizing towards energy efficiency and thermal comfort [3, 11]. Such building energy analysis distributes energy performances based on calculations from data.

The idea behind building energy analysis is to economically allocate annual energy budgets, economic optimization of energy, evaluating complicate with

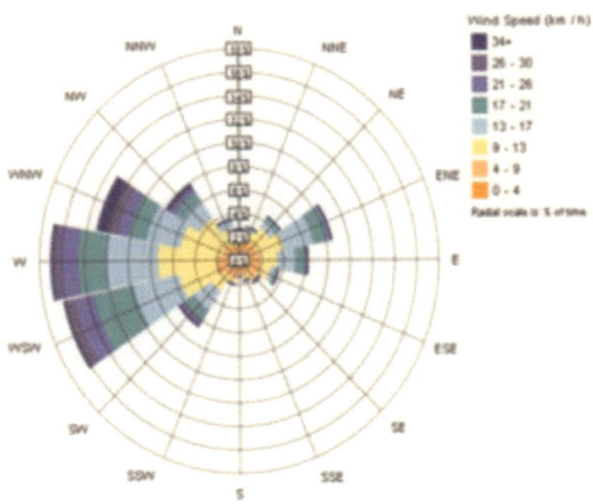

Figure 5.
Annual wind rose (speed distribution) monthly fuel consumption.

statutory energy standards and assessment of alternative components, systems and subsystems designs (see **Figures 3** and **4**). Notwithstanding the benefits of building energy analysis, there are normative issues bothering on guidelines and standards required in carrying out this processes as required by the European Union building energy performance guidelines who stipulated the methodology for carrying out building energy performance calculations [3]. The methodology requires a comprehensive analysis on heating installations, air-conditioning installation, positioning and orientation of building, natural ventilation, internal climate conditions, passive solar systems and solar protection, thermal characteristics of the building, Built-in lightning installation. It is often required that energy audit is conducted at post occupancy stage with the aim of addressing the deficits between initial design values to actual values at occupancy [3, 11, 12].

There are legislations amongst European nations addressing issues that are intrinsically related to ethical standards and practices. With reference to the above legislation, building energy analysis must as a matter of uniformity of practice, accommodate the following input data as requisite of an energy analysis process. These includes, utility rates, weather data, building orientation, thermal properties of building elements, building geometry and anthropometrics, building orientation in space, building energy load, and heating, ventilation and air-conditioning system [11–13].

The process of energy analysis runs building performance simulations with the aim of optimizing energy efficiency and too promotes carbon zeroing estimated in initial design stage (see **Figures 4** and 5). The economics of this process is well validated in its time and cost effectiveness in achieving high performance buildings. The general procedure for building energy analysis using AUTODESK Revit™ as copiously stated by Jangalve et al. [3] is that

1.Input BIM data for analysis

i. Project information

ii. Energy settings

iii. Materials

2. Input rooms/spaces/zones

   i. Define space limits

3. Define analysis information

   i. Reports

   ii. Schedule data

   iii. Details

4. Run heating and cooling load analysis

5. Export to gbxml for Autodesk GBS

6. Run or perform energy simulation

A typical output of building energy analysis illustrated in Jangalve et al., is shown in the accompanying graphical illustration in terms of CO_2 emission within the building arising from energy consumption (see **Figures 3** and **4**).

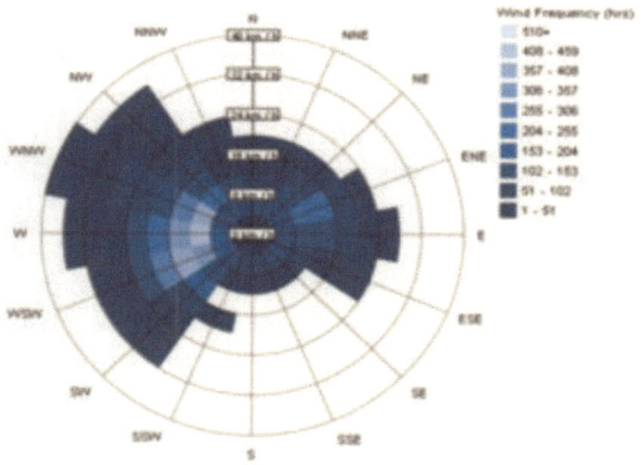

Figure 6.
Annual wind rose (frequency distribution) annual wind rose and humidity.

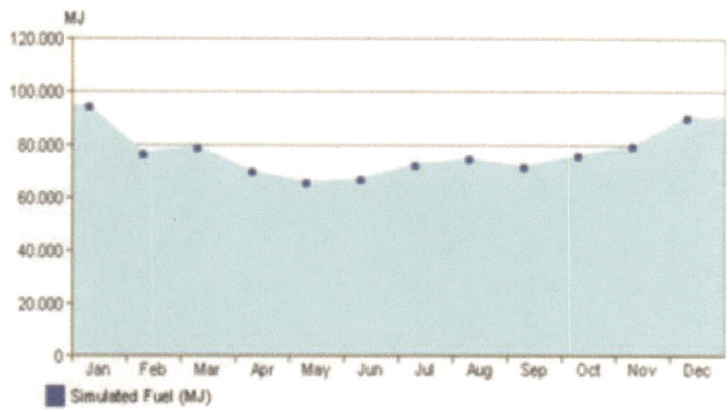

Figure 7.
Monthly fuel consumption.

In all, the use of energy simulation software to do analysis of building energy analysis is econometrically efficient in reporting the above listed evaluation outcome supports designers in actualizing building envelope properties and building energy requirements (**Figures 6** and 7).

4. Building energy audit

Attempts towards optimizing energy use in buildings stems from building energy analysis which in itself have been discussed extensively in previous section (**Figures 1–8**). Now for the possible of energy improvements, savings and optimization in a building which has its toot in building energy audit as a process requires the adaptation of agreeable and validated reference climates and energy simulation procedures [13, 14]. Such procedures look out for priority of energy uses in buildings. Energy audit as tool is used to benchmark where and how energy is being used in buildings. This is done by intensifying opportunities and providing solutions towards energy savings and economic cost too. Principal to this solution is energy data management suitable or compactable with appropriate energy saving technologies. It may also be in the form of structural improvements and systems modernization towards conservation of energy. In the process of energy auditing, the unified Lider Calener tool becomes indispensable to the process of auditing since it processing of results outcome allows for alternative response priorities to be earmarked [11, 15]. In economic sense, this process will serve as a guideline to developers and energy managers in taking decisions towards necessary reforms that will lead to substantial savings and enhanced payback period. In construction economics, the idea of conducting energy audit so as to save substantially ion energy is gaining sufficient space in construction literature and amongst developers, since the tool presents them the opportunity of making decisions on savings on energy consumption pattern with respect to economic indices of payback terms of the property investment [4]. The energy audit process seeks to advance and canvass improvement in the extent to which an energy budget is used for the intended purpose and cultivate energy savings towards offsetting greenhouse gas emission. It is important that energy auditing on the thermal properties of building envelopes should as a matter of economic science be conducted to give a closer direction of energy use to bring about decision bothering on the reasonability of a proposed investment [3, 15].

In as much as thermal comfort is important for occupants and must be paramount in design considerations, investors are usually worried about the cost

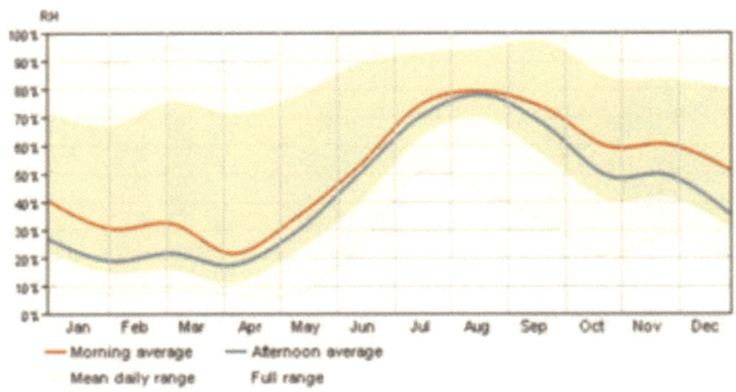

Figure 8.
Annual wind rose and humidity.

implication of such design been incorporate to meeting occupants satisfaction, yet must be comparatively admissible to be economically worthy to invest in. In the auditing process, thermal properties within the building envelope are assessed to ascertain the losses and gains that occur using the unified Lider Calener tool towards selecting the building element that will result to improved energy savings [16]. The decision to invest as an economic yardstick is to some extent dependent on the energy audit outcome which as a matter of necessity must take into cognizance the auditing method by considering time, speed, technical know-how, cost, sensitivity, accuracy, reproductively and ease of use. It is well articulated in building energy literature that buildings consume nearly one-third of the energy used in the United States [17]. This is not different if not more in most European nations. The operational cost of most buildings consequently absorbs at least 30% of operating cost. With this hindsight in the mind of developers, a constructability balance between economic gain and occupants comfort hangs on a balance that requires an economic assessment of cost-benefit appraisal by energy auditing. Particularly, building energy audit tends to reduce greenhouse gas emission and air pollution if properly done. It also addresses the air quality, lighting quality and occupants' satisfaction. It significantly lower electrical, natural gas, steam, water and sewer cost on the long run [13, 15]. Therefore, it becomes absolutely imperative that a knowledge bank of energy footprint of cities is established so as to identify the gaps of opportunities to savings in energy use and costs. The footprint repository will provide the necessary guidance for investors to making cost-benefit decisions, for the now and in the near future on energy saving alternatives and strategies [11, 13, 16].

5. Energy audit procedure

Before 2002, energy audit in the construction industry was not popular as what was available that somewhat looks like it was the European Union directive on energy efficiency of buildings, which was the 2002/91/EC, European Directive. In a later amendment of that document which was the 2006/32/EC, item 18 of the explanatory note mandates all member states to guarantee the availability of energy audit as statutory requirement and obligations in building construction projects. Subsequently in 2010 in 2010/31/EU specified the issuance of certificate of energy audit to the property owner as a mark of building's energy efficiency certification. Further to the provision of the 2010 document, the 2012 version made it obligatory and a routine of every 4 years activity for large companies with the attendant energy savings obtained in the period under review and to be inclusive of non-SME not later than December of 2015. In addition to the provisions of those directives, it stipulates that energy audit must show detailed calculation and proposed measure by furnishing clear information towards mitigating potential losses [11, 13, 15]. Since a rightly performed energy audit spells out the value of gain or loss at each energy point over a certain period, it will in time to come become a fundamental tool or document for showing compliance when energy intervention measures in a building that may lead to certain levels off savings in energy consumptions and reducing CO_2 emissions into the atmosphere are proposed. The efforts emanating from the accompanying legislations above are efforts geared towards minimizing non-renewable primary energy. Windows in buildings have been reported to be great sources of about 20–40% cause of energy losses in buildings, such that it must be taken into account when proposing energy-saving measures. Response measures towards the inhibition of energy uses via condensations must be considered from the view of avoiding a later investigation that could hamper the smooth running of thermal envelop elements [12, 15].

According to the New Jersey Energy Audit Guidance, there are three types of building energy audit which exists, namely

1. Computer simulation audit

This method is used to predict building systems performance taking external factors like weather into consideration. It is suitable for complicated buildings systems and facilities.

2. Walk-through audit

This method of building audit uses visual inspection of a building's energy system and review of its energy data usage. It is a further referenced by way of comparison to industry's normative average. This form of auditing determines if a further comprehensive audit is required. It is at best informative.

3. Standard audit

This method of building auditing is used to assess all equipment and the associated operational systems and generate more elaborate calculations of energy use. It primarily identifies areas of potential technical improvements and makes recommendations based on their projected energy and cost savings.

A typical energy audit procedure is enumerated hereunder.

- Collection of building data.
- Introduction of technical data into the HULC simulation program.
- Energy simulation of the building in its current state.
- Quantification of losses and gains through the different elements of the thermal envelop.
- Pre-selection of action measures based on the previous results.
- Simulation of possible refurbishments.
- Quantification of savings in each of the refurbishments proposed.
- Economic calculation and calculation of paybacks.
- Selection of final measures, based on the previous results.

Though the procedure is not limited to the arrangement stated above since it varies from practice to practice and from country to country, most European nations now have legislation and practice procedure. In Spain, for example, the Ministry of Development and the Ministry of Industry, Energy and Tourism freely issues the Unified Tool Lider Calener energy audit simulation program to practitioners to comply with national regulatory framework on energy performance of buildings for the purpose of energy certification. Typically, energy audit process is shown in the flow chart below:

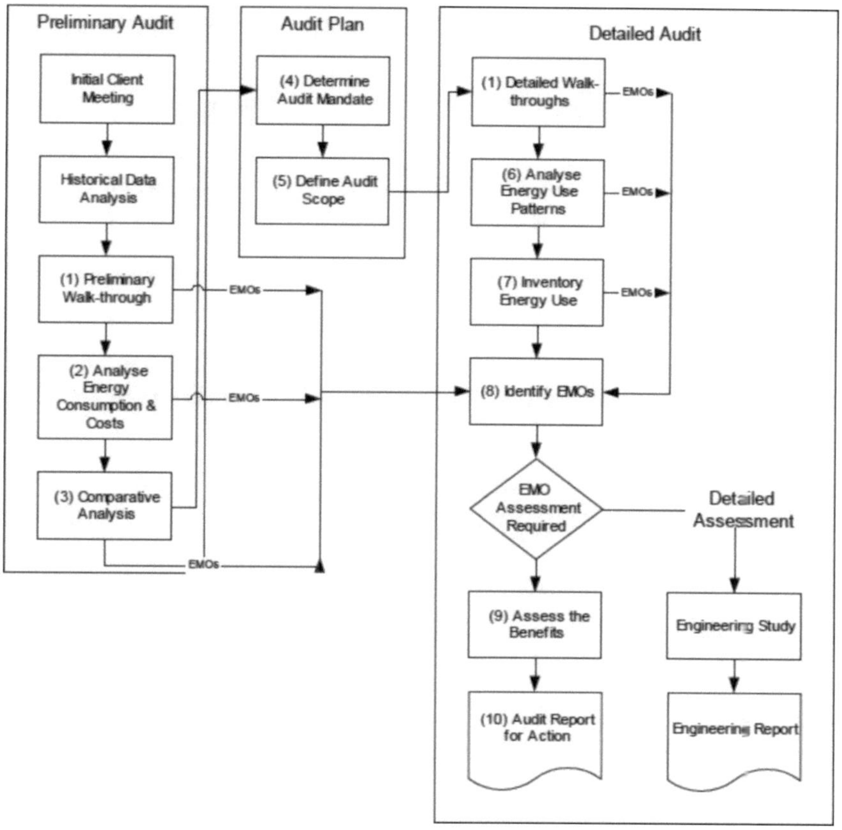

Energy audit flow chart
Source: South African Energy Department.

According to South African Energy Department, the key steps in the energy audit, after the initial client meetings and historical data analysis,

1. **Conduct a walk-through inspection**—to assess the general level of repair, housekeeping, and operational practices that have a bearing on energy efficiency, and to flag situations that have merit for further assessment as the audit is implemented, walk-through inspections will also be carried out to verify the findings of other analysis steps, as indicated in the flow chart;

2. **Analyze energy consumption and costs**—collect, organize, summarize and analyze historical energy billings and the tariffs that apply to them;

3. **Compare energy performance**—determine energy use indices and compare them internally from one period to another, one facility to a similar one within your portfolio, one system to a similar one; or externally to measures of good practice within your industry;

4. **Establish the audit mandate**—secure commitment from management and define expectations and outcomes of the detailed audit;

5. **Establish the audit scope**—define the energy consuming system to be audited.

6. **Profile energy use patterns**—determine the time relationships of energy use, as in the electricity demand profile;

7. **Inventory energy use**—prepare a list of all energy consuming loads in the audit area, and quantify their consumption and demand characteristics;

8. **Identify energy management opportunities**—including operational and technological measures to reduce energy waste.

9. **Assess the benefits**—quantify the level of energy and cost savings, along with any co-benefits.

10. **Report for action**—report the audit findings and communicate as required for implementation.

Each step involves a number of tasks that are described in the following sections. As suggested by the flow chart, several of the steps may result in the identification of potential EMOs.

6. Economics of building energy audit

Patel [4] did an extensive investigation on this subject matter. Patel [4] work noted that the auditor of the completed project has to be very careful in carrying out the audit. He must follow some procedure so that full justice is done to the work. Some points related to the energy audit procedure are described below.

6.1 Collection of appropriate information

The starting point for collecting post audit information is the project completion report. Energy audit generally compares the projected data with the accounting data collected through the regular MIS. The MIS needs to be geared up so that the projected cash flow from the original capital budget can be compared with the actual cash flows realized during the period elapsed before the energy audit of the project has started. Another point that needs to be kept in mind is that the auditor needs to collect total cost figures. Incremental cash flow figures are readily available for green-filled projects but it is not so easy for the projects in an existing plant. The data in the latter case need to be appropriately dealt with to arrive at the incremental cost figures due to [4].

6.2 Recasting the data

Collected data or budgeted data should be recast before they are compared. The significant time gap and a host of factors, which were not considered at the budgeting stage, would warrant the recasting of data. For example, inflation should be adjusted before comparison is made. Sales mix difference due to external factors also should be taken into account. In the absence of adjustment for those "external" factors, the quality of audit would suffer. Inflation adjustment is subsequently into results.

6.3 Comparison of projected financial parameters with actual

This is the next important step in the post completion audit procedure. There are four techniques available for the comparison of actual with the projected financial parameters. The comparison is the starting point from which the real audit begins. Only comparable data is compared. Adjustments are first done for inflation and external factors before comparison is carried out under any method. Methods described later on are not mutually exclusive. More than one method may be applied for comparison if there is such a requirement. A broad level ROI or NPV comparison can be done initially, followed by detailed cost variance or cash flow variance analysis. Comparison is a step-by-step approach so that causes are identified systematically with minimum cost, time and energy [4].

$$ROI = \frac{Net\ \text{income}}{\text{Cost of Investment}}$$

$$= \frac{\text{Investment gain}}{\text{Investment base}}$$

$$= \frac{\text{Gain from investment} - \text{cost of investment}}{\text{Cost of Investment}}$$

$$= \frac{\text{Return(Benefit)}}{\text{Investment (cost)}}$$

6.4 Establish the possible causes of variance

Once the variance figures are calculated, if they are significant, the possible causes for the same are explored. An auditor goes by exceptions from there he tries to reach the root causes of deviations. This process of investigation can be effective only if an auditor possesses skills of inquisitiveness and skills of persuasion and negotiation [4]. A summary report of the energy audit findings should also be prepared.

6.5 Final recommendations

Once the causes are ascertained, the post completion auditor can give his recommendations based on which the manager may take decisions for cash flow forecasting to reinvest or abandon the ongoing project. Hopefully, after the post completion audit, the cash flow prediction and project evaluation become more accurate [4].

7. Building energy audit economic techniques

There are three techniques of building energy audit economics, namely

a. Cost variance analysis,

b. Profit variance analysis

c. Cash flow and financial criteria analysis and

d. Present value depreciation technique.

7.1 Cost variance analysis

In this method, only the project cost (actual and estimated) is studied and the revenue aspect is not included in the audit. This approach is adopted when the energy audit is conducted during the execution or just after the completion of the project.

CV [Earned Value (EV) – Actual Cost (AC)]
EV = % of worth completed x Budgeted cost
AC = what has been spent on the project

7.2 Profit variance analysis

In this method, plant-wise profit analysis is carried out by the auditor and the estimated gain adjusted with the inflationary effect is compared with actual. An important point to note here is that even if the aggregates of gains (realized and estimated) are the same there can be wide variations for individual projects, indicating the need for further investigations

7.3 Cash flow and financial criteria analysis

This method is developed around four schedules described below. These schedules can provide the management with the information it needs to find engineering, operational and administrative costing faults of past projects.

a. Profit variance analysis schedule: this schedule is prepared for the calculation of profit variance between projected and the actual project results. The information for the "projected" column is obtained from the approved capital expenditure request. The information for the "actual" column is obtained from regular accounting sources. Supplementary schedules are required to itemize and explain the basis of calculation of revenues, costs and expenses need to be given.

b. Cash flow and financial criteria variance analysis schedule: this is used to illustrate project cash flow and return variances between the projected and actual results. The approved capital expenditure request is again used to provide information for the "projected" column and regular accounting sources for the "actual" column.

c. Project cash flow schedule (projected and actual): these are used to show the projected and actual cash flows of the project. They illustrate the timing of cash flows to compute payback and to provide the net period cash flow information required for the IRR calculation. Each cash flow entry is made according to the time it was projected to be incurred or was actually incurred. The period cash flows are for individual quarters whereas the cumulative cash flows represent all cash flow for the project. The payback point is reached when the cumulative net cash flow equals zero.

d. Supplementary schedules: the supplementary schedules provide explanations for the significant variances.

7.4 Present value depreciation technique

Discounting factor technique give only a single value of the NPV which is for the whole life of the project. The IRR is the average return during the life. But at the time of conducting the energy audit, the major part of the project life is not completed. Then how can we compare the actual with the total net present value or average internal rate of return? A uniform annual series cannot be considered because it is an average figure and the project need not offer and NPV at a constant rate over its life. The concept of e present value of depreciation is used in some techniques for the calculation of the year-wise NPV and IRR. Present value depreciation is defined as the decline in the present value of the expected future cash flow during the year using the IRR as the discount rate. Two models, namely the IRR model and the NPV model, are suggested under the technique of present value depreciation in Eq. (51).

$$NPV = \sum_{t-1}^{n} \frac{CF_t}{(1-K)^t} - C \tag{51}$$

where CF_t = cash flow in period t; K = discount rate; C = initial outlay

8. Conclusions

There are numerous literatures that have dealt with this subject of building energy audit. This particular text emphasizes economic assessment moving from a theoretical review of the subject of building energy analysis with respect to thermal conditions of the building envelope. The process of obtaining data for building energy audit was spelt out from the review of prevalent building information modeling software which were analyzed to obtain actual and project energy loads. The point of divergence of the two measures were econometrically reviewed to ascertain financial control mechanism, providing information for future capital expenditure decision, impacts on proposals for capital investments.

Acknowledgements

I am indebted of thanks and gratitude to authors whose text materials were used to form the theoretical bedrock for economic analysis particularly of mention is the Kreider J.F. and Rabl A., Jangalve, A., Kamble, V., Gawandi, S., and Ramani, N. and Patel, B.M. I cannot thank them enough.

Conflict of interest

This text has no conflict of interest declaration.

Author details

Samuel I. Egwunatum[1*] and Ovie I. Akpokodje[2]

1 Department of Quantity Surveying, Federal University of Technology, Owerri, Nigeria

2 Department of Civil Engineering, Delta State Polytechnic, Ozoro

*Address all correspondence to: samuelegwunatum@gmail.com

References

[1] Kreider JF, Rabl A. Heating and Cooling of Buildings: Design for Efficiency. New York: McGraw Hill Inc; 1994

[2] Fine HA, Jury SH, Yarbrough DW, McElroy. Analysis of Heat Transfer in Building Thermal Insulation. Department of Energy Conservation and Solar Energy Office of Buildings and Community System, US Department of Energy; 1975

[3] Jangalve A, Kamble V, Gawandi S, Ramani N. Energy analysis of residential building using BIM. International Journal of Emerging Engineering Technology and Science. 2017;**108**:15-19. Available from: http://www.ijeets.org

[4] Patel BM. Project Management: Strategic Financial Planning, Evaluation and Control. New Delhi: Vikas Publish House PVT Ltd; 2012

[5] Sparrow EM, Cess RD. Radiation Heat Transfer. Belmont, Calif: Brooks/Cole Publishing Co; 1970

[6] Viskanta R. Heat Transfer in Thermal Radiation absorbing and Scattering Media. 1960. ANL-6170

[7] Viskanta R, Grosh RJ. Effect of surface emissivity on heat transfer by simultaneous conduction and radiation. International Journal of Heat and Mass Transfer. 1962;**5**(8):729-734

[8] Rennex BG. Thermal parameters as a function of thickness for combined radiation and conduction heat transfer in low-density insulation. Journal of Thermal Insulation. Journal of Building Physics. 1979;**3**(1):37-61. DOI: 10.1177/109719637900300104

[9] Heaslet MA, Warming RF. Radiative transport and wall temperature slip in an absorbing planar medium. International Journal of Heat and Mass Transfer. 1965;**8**:979

[10] Siegel R, Howell JR. Thermal Radiation Heat Transfer. New York: McGraw-Hill; 1972

[11] Commission Recommendation (EU) 2016/1318 of 29 July 2016 on guidelines for the promotion of nearly zero-energy buildings and best practices to ensure that, by 2020, all new buildings are nearly zero-energy buildings (OJ L 208, 2-August-2016, p. 46)

[12] Directive 2010/31/EU of the European Parliament and of the Council of 19 May 2010 on the energy performance of buildings (OJ L 153, 18-June-2010, p. 13)

[13] Directive 2009/148/EC of the European Parliament and of the Council of 30 November 2009 on the protection of workers from the risks related to exposure to asbestos at work (OJ L 330, 16-December-2009, p. 28)

[14] Directive (EU) 2016/2284 of the European Parliament and of the Council of 14 December 2016 on the reduction of national emissions of certain atmospheric pollutants, amending Directive 2003/35/EC and repealing Directive 2001/81/EC (OJ L 344, 17-December-2016, p. 1)

[15] Directive 2012/27/EU of the European Parliament and of the Council of 25 October 2012 on energy efficiency, amending Directives 2009/125/EC and 2010/30/EU and repealing Directives 2004/8/EC and 2006/32/EC (OJ L 315, 14-November-2012, p. 1)

[16] Regulation (EU) No 182/2011 of the European Parliament and of the Council of 16 February 2011 laying down the rules and general principles concerning mechanisms for control by the Member

States of the Commission's exercise of implementing powers (OJ L 55, 28-February-2011, p. 13)

[17] Viskanta R. Heat transfer by conduction and radiation in absorbing and scattering materials. Journal of Heat Transfer. 1965;**87C**:143